D0434859

Wonderful Life the Elements

The Periodic Table Personified

by Bunpei Yorifuji

WONDERFU

Wonderful
by Kagaku
arranged

All rights
any means
storage or
publisher.

Printed in

First print

15 14 13

ISBN-10:
ISBN-13:

For inform

No Starch
38 Ringold
phone: 41

Library of
A catalog

No Starch
Other proc
owners. Ra
are using
no intentic

The inforn
precaution
Press, Inc.
caused or

All charact
dead, is pu

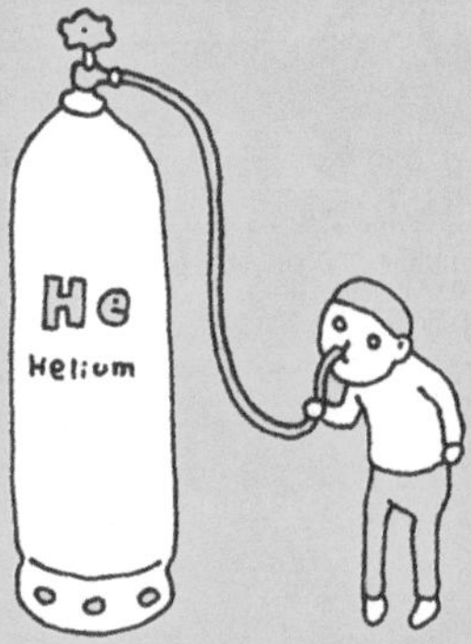
He
Helium

PREFACE

Do you know what happens if you inhale a lot of helium? Back when I was an art student, I bought two canisters of pure helium for one of my works. Inhaling helium, as you might know, raises the pitch of your voice. But common helium balloons don't really raise your voice that much, and it goes back to normal right away.

"BUT I MIGHT BE ABLE TO PRODUCE SOME REALLY FUNKY NOISES WITH THESE."

So I exhaled with all my might, opened one of the canisters, and filled my lungs with as much helium as I could. And everything just went black. I tried to breathe, but all I could really do was gasp, as no air would grace my lungs. I could feel the warmth leaving my body as I started to lose consciousness. It was only after this experience that I learned that inhaling pure helium can lead to suffocation and death.

Since I was all alone in the lab, I decided it might be a good idea to call out for help.

IN SUPER SOPRANO: "HELP MEEE...."

But that voice! Inhaling helium is dangerous in more than one way. The first is that it suffocates you, and the second is that even if you call for help, your cries will probably be dismissed as a bad practical joke.

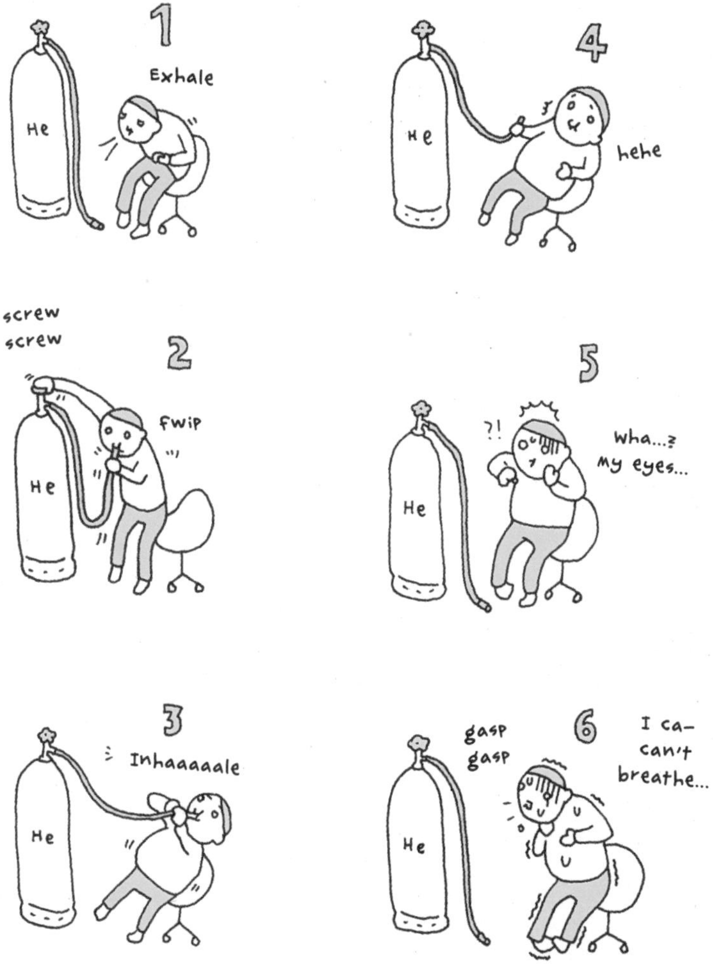
1
Exhale
screw
screw
2
fwip
3
Inhaaaaale
4
hehe
5
?!
Wha...?
My eyes...
6
gasp
gasp
I ca-
can't
breathe...

We're usually not aware of the elements in our daily lives. We don't look at a desk and instantly think "Carbon!" And knowing a lot about the elements doesn't really make you cool (in fact it's quite the opposite).

THE CONCEPT OF ELEMENTS DOESN'T COME NATURALLY TO US.

First of all, protons, neutrons, and electrons are all so small. And the idea that you can split this complex world into 118 basic elements isn't easy to believe. But the concept of the elements also has this aura of serenity that is hard to resist—a promise that hints at the true core of all matter. However, they are still too small to for us to care about in our daily lives, and they're too abstract to serve as explanations for why the things around us are as they are.

In this book, I've tried to distill these seemingly abstract little things into something that might be easier to grasp. This book was written with the help and supervision of Kouhei Tamao of the Institute of Physical and Chemical Research, Hiromu Sakurai of Kyoto Pharmaceutical University, and Takahito Terashima of Kyoto University. I don't think there is any real point in trying to remember everything about every element, but I hope that you'll learn a little about each and every one of them—and have fun—by reading this book.

Help meee...
He
surprised at
his own voice

But it went back to
normal right away.

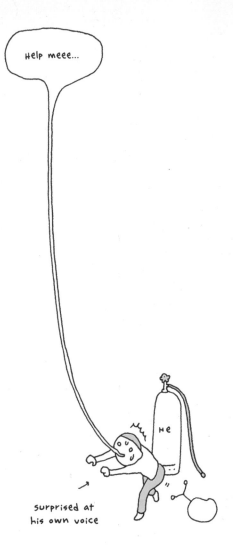

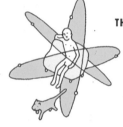

元素の食べ方
HOW TO EAT THE ELEMENTS

p.171

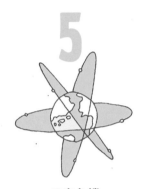

元素危機
THE ELEMENTS CRISIS

p.195

CONTENTS

1

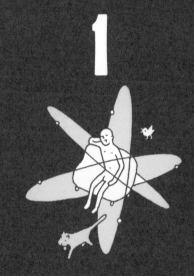

ELEMENTS IN THE LIVING ROOM

リビングと元素

宇宙を構成する元素

ELEMENTS OF THE UNIVERSE

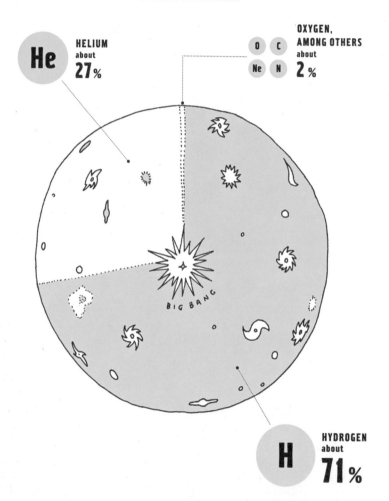

He HELIUM
about
27%

OXYGEN,
AMONG OTHERS
about
2%

O C
Ne N

H HYDROGEN
about
71%

太陽を構成する元素

ELEMENTS OF THE SUN

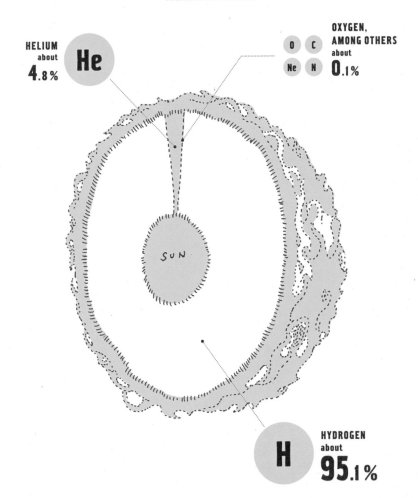

HELIUM
about
4.8%

He

OXYGEN,
AMONG OTHERS
about
0.1%

O   C

Ne   N

SUN

HYDROGEN
about
95.1%

H

地球を構成する元素

ELEMENTS OF EARTH

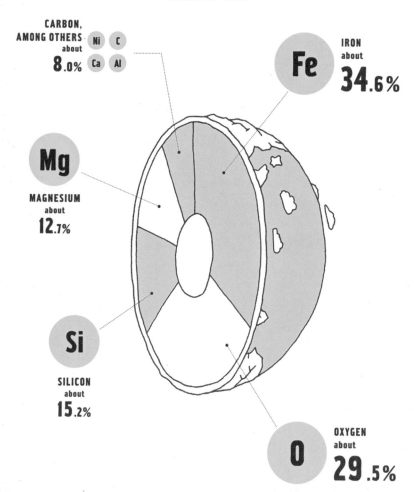

CARBON, AMONG OTHERS about **8.0%**
Ni C
Ca Al

IRON about **34.6%**
Fe

MAGNESIUM about **12.7%**
Mg

SILICON about **15.2%**
Si

OXYGEN about **29.5%**
O

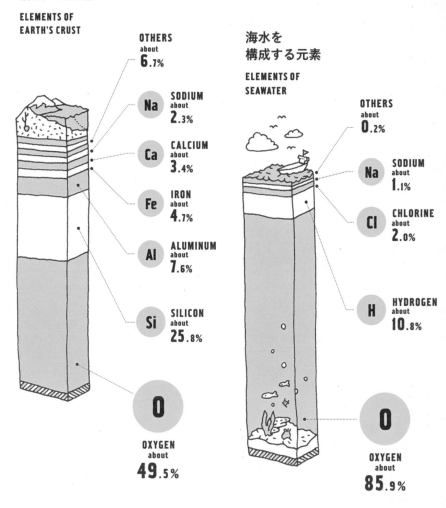

地殻を
構成する元素

ELEMENTS OF
EARTH'S CRUST

OTHERS
about
6.7%

Na SODIUM
about
2.3%

Ca CALCIUM
about
3.4%

Fe IRON
about
4.7%

Al ALUMINUM
about
7.6%

Si SILICON
about
25.8%

O OXYGEN
about
49.5%

海水を
構成する元素

ELEMENTS OF
SEAWATER

OTHERS
about
0.2%

Na SODIUM
about
1.1%

Cl CHLORINE
about
2.0%

H HYDROGEN
about
10.8%

O OXYGEN
about
85.9%

Elements fit perfectly in discussions of things like planets and outer space. But discussing our daily lives from the perspective of elements usually doesn't make much sense. In the last billion years or so, the elements of Earth haven't changed much. And it doesn't matter to the elements whether people live or die—it's all the same to them.

ENVIRONMENTAL PROBLEMS DON'T AFFECT THEM EITHER.

The elements remain unaffected even if holes open up in the ozone layer or the atmosphere fills up with carbon dioxide. Unless something really drastic happens, like a meteor strike or a nuclear bomb, there's really no change to the elements of Earth. But if something like that happens, then nothing really matters anymore, does it? It becomes hard to even start comparing our daily lives to the lives of the elements when we think about it like this.

But even though there's no change in the elements themselves, if we look at a time span of say 10,000 years, a change in the way we use the elements can clearly be seen. Let's take a look at that next.

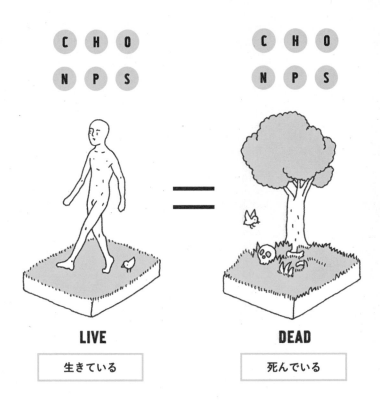

LIVE

生きている

DEAD

死んでいる

No change

H
O
P
C
Si
N
S

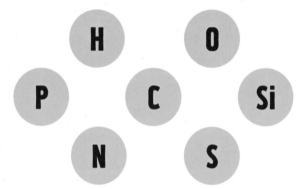

WOOD AND GRASS

C H O

N P S

Si O

SOIL AND DIRT

原始の生活

PRIMITIVE TIMES

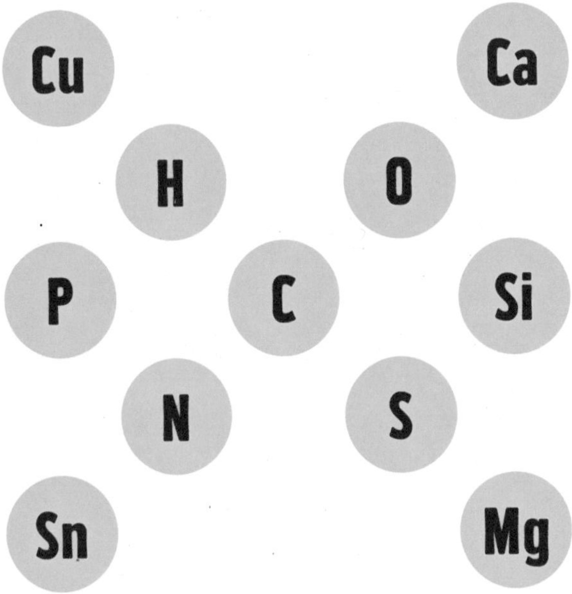
Cu
Ca
H
O
P
C
Si
N
S
Sn
Mg

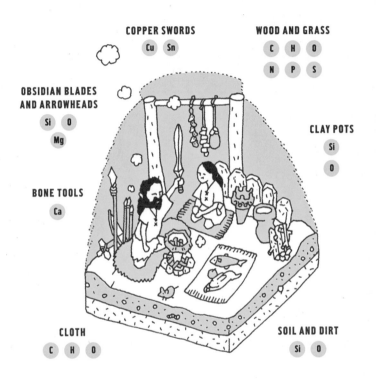

COPPER SWORDS
Cu Sn

WOOD AND GRASS
C H O
N P S

OBSIDIAN BLADES
AND ARROWHEADS
Si O
Mg

CLAY POTS
Si
O

BONE TOOLS
Ca

CLOTH
C H O

SOIL AND DIRT
Si O

古代の生活

ANCIENT TIMES

Al
Fe
Au
H
O
P
C
Si
N
S
Cu
Ca
Mg
Ag
Co

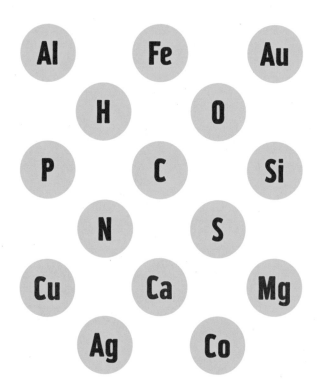

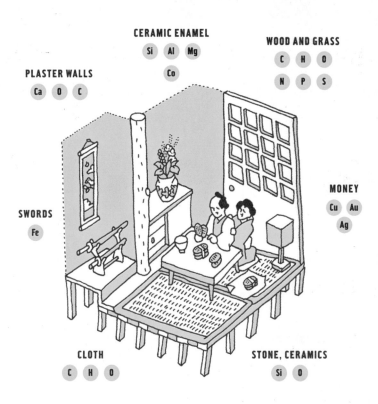

CERAMIC ENAMEL
Si Al Mg
Co

PLASTER WALLS
Ca O C

WOOD AND GRASS
C H O
N P S

SWORDS
Fe

MONEY
Cu Au
Ag

CLOTH
C H O

STONE, CERAMICS
Si O

中世の生活

MEDIEVAL TIMES

As Fe W

Ru Zr In Sb

Al Cu Au

Ga H O Nd

P C Si

Li N S Hg

Br Ag Ni

Mn Co Ta Te

Mo Pb Kr

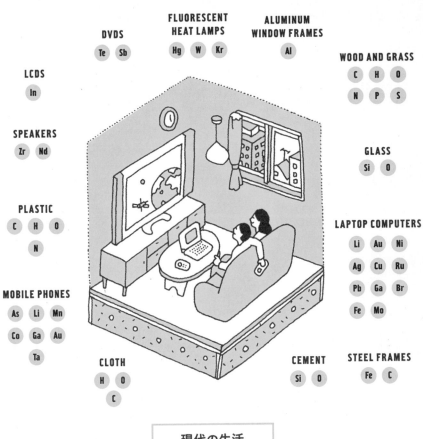

DVDS
Te Sb

FLUORESCENT
HEAT LAMPS
Hg W Kr

ALUMINUM
WINDOW FRAMES
Al

LCDS
In

WOOD AND GRASS
C H O
N P S

SPEAKERS
Zr Nd

GLASS
Si O

PLASTIC
C H O
N

LAPTOP COMPUTERS
Li Au Ni
Ag Cu Ru
Pb Ga Br
Fe Mo

MOBILE PHONES
As Li Mn
Co Ga Au
Ta

CLOTH
H O
C

CEMENT
Si O

STEEL FRAMES
Fe C

現代の生活

TODAY

The number of elements we use every day has been steadily increasing over the last 10,000 years, with an especially sharp increase over the last 50 years or so. We use five times more elements than in primitive times and twice as many as in medieval times.

ELEMENTS FROM ALL CORNERS OF THE WORLD GATHER IN OUR LIVING ROOMS.

The indium used in our LCD TVs is from China, and plastic and vinyl are from oil drilled in the Middle East. (Oil is made up of carbon, mind you.) With the recent spread of the Internet, our borders have opened up with the help of copper and silicon dioxide (the elements that make up fiber-optic cables). Just imagine all the photons and electrons flying around the world. It probably wouldn't be a lie to say that this is the first time since the last cataclysmic asteroid struck Earth that this many different elements are being used at the same time.

When we say "global," most people think of the economy, or maybe politics. But there is probably nothing as "global" as the basic elements. We are always connected to the rest of the world through the elements in our technology.

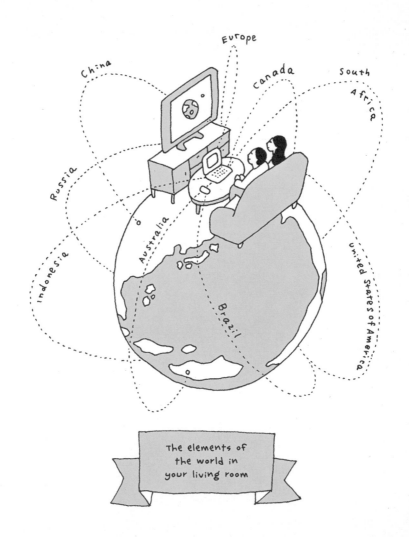

The elements of
the world in
your living room

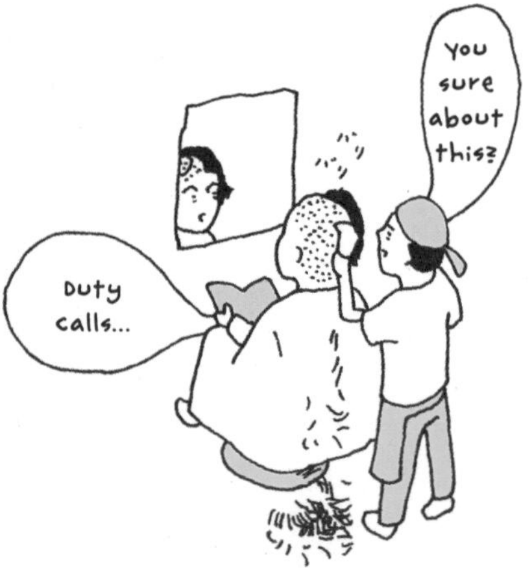
You sure about this?
Duty calls...

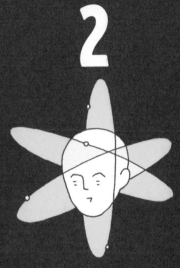

2

THE SUPER PERIODIC TABLE

OF THE ELEMENTS

スーパー元素周期表

元素周期表

THE PERIODIC TABLE OF THE ELEMENTS

Basic elements are usually represented using letters, like *F* and *H*. The rows in the table are called *periods*, and the columns are called *families* or *groups*. Since there are so many elements in both the Ln and An families, they've been given their own space at the bottom. Understanding the structure of the periodic table can really help when trying to learn about the amazing world of the elements.

PERIOD \ FAMILY	1	2	3	4	5	6	7	8	9
1	H								
2	Li	Be							
3	Na	Mg							
4	K	Ca	Sc	Ti	V	Cr	Mn	Fe	Co
5	Rb	Sr	Y	Zr	Nb	Mo	Tc	Ru	Rh
6	Cs	Ba	Ln	Hf	Ta	W	Re	Os	Ir
7	Fr	Ra	An	Rf	Db	Sg	Bh	Hs	Mt

Ln = La Ce Pr Nd Pm Sm Eu

An = Ac Th Pa U Np Pu Am

								He
		B	C	N	O	F	Ne	
		Al	Si	P	S	Cl	Ar	
Ni	Cu	Zn	Ga	Ge	As	Se	Br	Kr
Pd	Ag	Cd	In	Sn	Sb	Te	I	Xe
Pt	Au	Hg	Tl	Pb	Bi	Po	At	Rn
Ds	Rg	Cn	Uut	Fl	Uup	Lv	Uus	Uuo
10	11	12	13	14	15	16	17	18

Gd	Tb	Dy	Ho	Er	Tm	Yb	Lu
Cm	Bk	Cf	Es	Fm	Md	No	Lr

HARRIET LIKES NAVY KARL'S RUBBER-COATED FRIGATE.

I'm sure many of you used nonsensical mnemonic tricks like this one to memorize the periodic table just like I did.

This is a pointless waste of time.

The elements were originally arranged in this way according to the number of protons present in the atomic core, but this number also determines the number of electrons orbiting the core, and this number in turn determines the behavior of the atom, which finally determines the atom's properties. "Harriet Likes Navy Karl's..." is only a simple memorization tool to help you learn the elements' names; it doesn't help you actually get to know them.

That's why we have the periodic table.

The periodic table is the amazing result of many scientists' knowledge and hard work. But even so, it doesn't make much sense the first time you see it. By making each element's properties obvious at a glance, I've created a periodic table that should be a bit more accessible to newcomers.

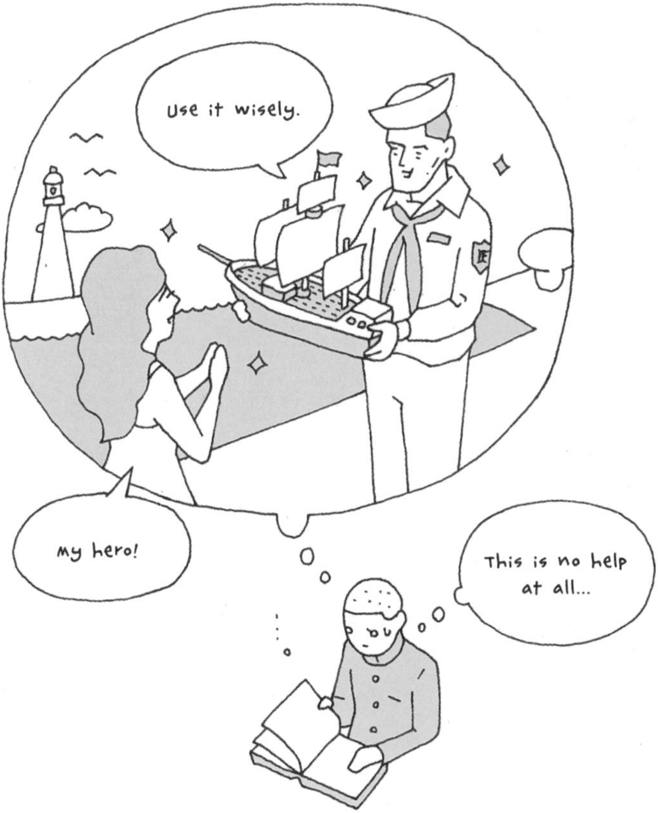
Use it wisely.
My hero!
This is no help at all...

通常の原子の表し方

ELEMENTARY PARTICLE NAMES

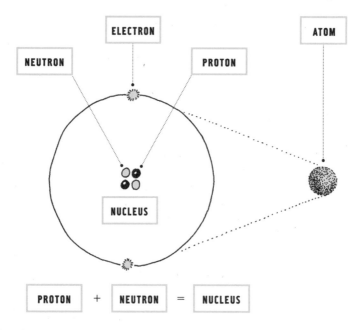

ELECTRON

NEUTRON

PROTON

ATOM

NUCLEUS

PROTON + NEUTRON = NUCLEUS

Atomic names are used to classify the basic elements.

Atoms are made up of a nucleus and orbiting electrons. The nucleus consists of two kinds of particles called *protons* and *neutrons*.

Protons and electrons are electrically charged; protons are positive and electrons are negative. An atom in its most basic form is electrically balanced, which means that there is an equal number of protons and electrons. If additional electrons are added or removed, we say that the atom becomes *ionized*, and it is consequently called an *ion*.

The electrons orbiting the nucleus move very fast and are therefore collectively called the *electron cloud*. I simplified the cloud in the drawing above so that individual electrons can be seen.

原子を顔で表す

THE ATOM AS A FACE

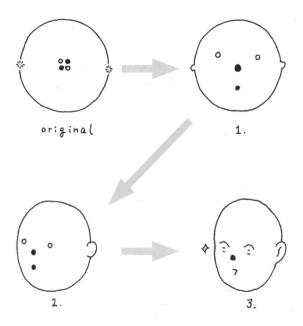

original

1.

2.

3.

Each electron belongs to an electron shell. As the number of electrons increases, new shells are formed farther away from the nucleus. The electrons belonging to the outer shell are called *valence electrons*. Interactions between atoms are governed by their valence electrons, and many atomic properties are derived from the number of these electrons.

As you can see, I rearranged this atom into a face: The neutrons became eyes, and the protons became the nose and mouth. While not exactly scientific, this presentation should make for a much more attractive collection of elements.

元素のヘアースタイル

Hairstyles of the elements

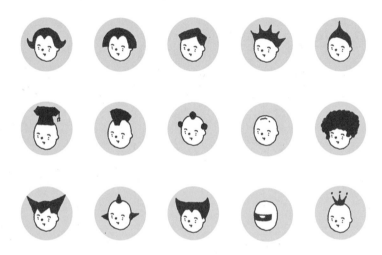

I've split the properties of the elements into 14 categories. (Hydrogen is in a class by itself.) They're mostly organized according to the families in the periodic table, but since some elements belonging to the same family exhibit different properties and elements of different families can be similar, I decided to alter these categorizations slightly. I tried to model each group's hairstyle after its chemical properties.

アルカリ金属

Alkali metals

Floaty, flirty hair.

All elements of the 1st family except hydrogen. They're very soft for being metals and can even be cut with a knife. They're also not very dense, so they float in water. And they oxidize easily, which means they quickly lose their luster.

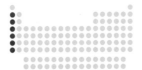

アルカリ土類金属

Alkaline earth metals

**A bit plain.
Pudding bowl cut.**

The metals belonging to the family in the lower part of the 2nd column from the left. They're highly reactive and can bind to the oxygen and moisture in the air, although not as easily as the alkali metals. They're commonly found in rock, hence the "earth" in the family name.

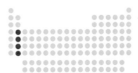

遷移金属

Transition metals

**The majority of the metal elements.
Clean-cut and boring.**

The elements from the 3rd to the 11th families. These are the multitude of elements usually referred to as *metals*. They all possess very similar properties, and there are a lot of them.

亜鉛族

The zinc family

**Volatile.
Punk hair.**

The four elements of the 12th family. Mercury is different from zinc and cadmium in that it's the only metal that's in liquid form at room temperature. These elements all evaporate easily, have low melting points, and are volatile.

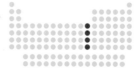

ホウ素族

The boron family

Light and sharp.
Pointy hair.

The elements of the 13th family. Aluminum is their front man, appearing in many modern applications. The family's name might rhyme with "moron," but don't underestimate these elements—gallium, indium, and the rest of them are all used in cutting-edge technology.

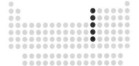

炭素族

The carbon family

The talented ones.
Intellihair.

The elements of the 14th family. Carbon is highly reactive, which means it will bind with many different elements and can be found in almost all organic compounds. Silicon is widely used as a semiconductor. Lead, germanium, and tin were very popular back in the day but don't make many appearances nowadays.

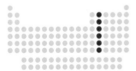

窒素族

The nitrogen family

**Hates normal.
Mohawk.**

The five elements in the 15th family. All of them are solids at room temperature except for nitrogen, which creates very stable molecules that make up about 80% of our atmosphere. Many of these have been known for ages, among them phosphorus and arsenic, which made good poisons among other things.

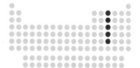

酸素族

The oxygen family

**Old school.
The half-assed bald shave.**

The 16th family, consisting of six elements. Oxygen is the only gas at room temperature. Sulfur, selenium, and tellurium are all ores and minerals that make up common rocks. Polonium is slightly radioactive. This group is often referred to as the *chalcogens.*

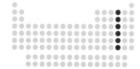

ハロゲン

Halogens

**Bald and bulbous,
like a halogen lamp.**

The nonmetallic elements of the 17th family. At room temperature, fluorine and chlorine are gases, iodine and astatine are solids, and bromine's a liquid, so they're not very similar in that respect. But they're all highly reactive and create salts when bound to elements from the alkali and alkaline earth families.

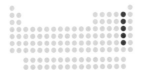

希ガス

Noble gases

**Too cool.
Afro.**

The six elements of the 18th family. They're the most stable elements of all and therefore seldom react. They all have low boiling and melting points. Helium doesn't solidify even at absolute zero (–273.15°C).

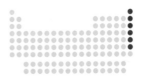

ランタノイド

Lanthanides

**Very rare.
Astro hair.**

The 15 elements starting with lanthanum and ending with lutetium. They are extremely rare and are therefore sometimes called the rare-earth elements. Some of them possess very similar properties and can be difficult to tell apart. It took over 100 years to find them all.

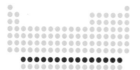

アクチノイド

Actinides

**Mostly man-made.
Robot hair.**

Actinides is the umbrella name for the 15 elements starting with actinium and ending with lawrencium. Their properties are very similar to the lanthanides series', and almost all of them are man-made. The elements after neptunium are all heavier than uranium, so they're sometimes called *transuranic*.

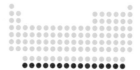

その他

Other metals

**The outsiders.
Weird hair.**

Beryllium and magnesium are in the same column as the alkaline earth metals, but I've decided to put them into their own category since they don't display some of the characteristics common to the others. For instance, they don't burn with any particular color when subjected to the flame test, while the other four do.

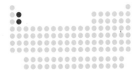

特別枠

Hydrogen and the Unun series

**The supreme ruler
and the shrouded unknowns.**

Hydrogen holds a special place in the universe, as it's the simplest element of them all but makes up roughly 71% of the known universe. The properties of the hard-to-remember unun series in the other corner of the table, however, are still more or less unknown.

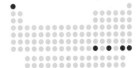

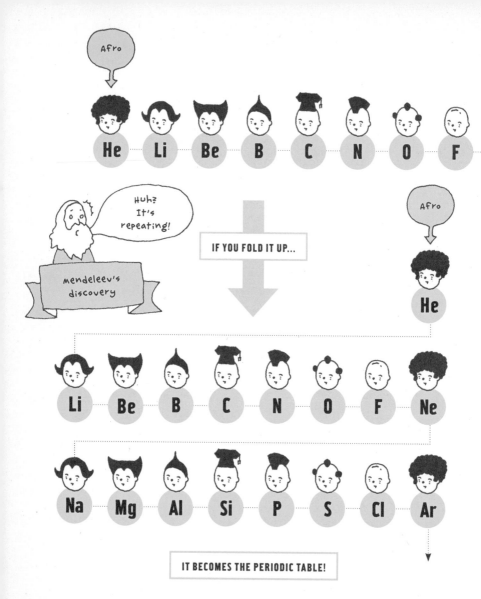

IT BECOMES THE PERIODIC TABLE!

Now that we've split the elements into categories, let's line them up and look for a pattern. Do you see it?

The elements, if arranged according to their atomic weight, exhibit an apparent periodicity of properties.

This is what the Russian scientist Dmitri Mendeleev discovered and wrote in his presentation "The dependence between the properties of the atomic weights of the elements." He pointed out that this periodicity can be used to create a table where elements of the same column exhibit similar properties, and get heavier with each row. This discovery eventually matured into the periodic table we know today.

Just because we managed to split the elements into different categories doesn't mean that they don't have their individual quirks and properties. Wouldn't it be great if we could make a periodic table where you could see all these properties right away, just by looking at each element? Something like a *super* periodic table of the elements...

固体・液体・気体をカラダで。

Matter states as body types

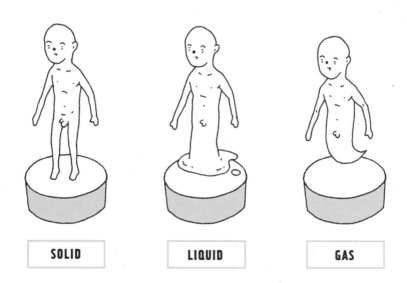

SOLID	LIQUID	GAS

Let's not stop at faces. Let's do their bodies too!

At room temperature, some elements (like iron) are solid, others (like mercury) are liquid, and yet others (like oxygen) are gaseous. I'm going to let the lower half of their bodies indicate which form they normally have. Gases will be ghosts, liquids will be aliens from Planet X, and solids will be humans. There are only two natural liquids though, so most of them will be solids or gases.

原子量を体重で。

Atomic weight as body weight

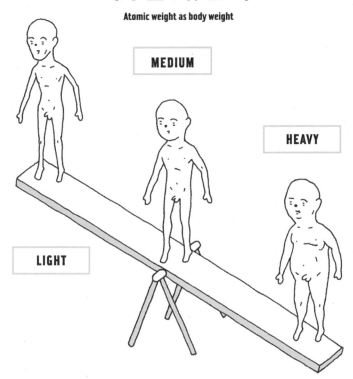

MEDIUM

HEAVY

LIGHT

One *atomic weight unit* is equal to one-twelfth of a carbon-12 atom's weight—but let's leave the technical stuff for another time. As you can see, I decided to model atomic weight as body weight. Atoms generally get heavier the farther you go in the periodic table, so my drawings will just keep getting fatter. It is worth noting that roentgenium (atomic number 111) is about 270 times as heavy as the lightest element, hydrogen. So instead of trying to model the exact relationships between the atoms, which would force me to draw the biggest elements several pages large, I'll just try to capture the general feeling of their relative sizes.

発見された年を年齢で。

Discovery year as age

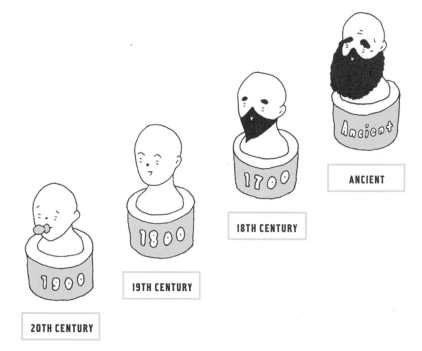

ANCIENT

18TH CENTURY

19TH CENTURY

20TH CENTURY

Some elements were discovered ages ago, and some synthetic ones were discovered only recently.

I thought I'd model their ages after how long we've known about them. Most elements were discovered during the 19th century, so using that as a baseline, I decided on these four simple categories.

特殊な性質は背景や服で。

Special properties as backgrounds and clothes

| **RADIOACTIVE** | **MAGNETIC** | **LUMINESCENT** |

Radioactive elements. They can be difficult to handle but have many important uses.

Elements that generate powerful magnetic fields. I decided on a fancy two-tone suit to match the duality of a magnet's north and south poles.

Elements used for luminous paint, fire-works, and fiber-optic cables.

I tried to make it extra clear which elements possess radioactive, magnetic, and luminescent properties. The mark around the radioactive character is inspired by the real radioactivity hazard symbol, which warns of alpha, beta, and gamma radiation.

    Magnetic elements will be easily recognized by their two-color suits.

The real mark looks like this.

おもな使用用途を服装で。

Usage areas as clothes

 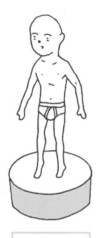 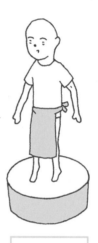 

MULTIPURPOSE	MINERAL	DAILY	INDUSTRIAL
These versatile team players are popular in most application areas.	Elements used by our bodies as nutrients are dressed to show off their healthy physique.	The nurturing materials we encounter every day in our kitchens and living rooms.	The businessman elements that work in our industries and factories.

Some elements are used by all of us, and some are used only by scientists. I decided to illustrate their applications by giving them different clothes, but it proved more difficult than I first anticipated. Some elements are used in many different areas, which makes it hard to say that they belong to any single one. But the categories should serve as a general pointer at least.

| SPECIALIST | SCIENTIFIC | MAN-MADE |

Elements used only in specialized applications wear coveralls.

Elements not yet used by the general public but that can be found in research laboratories wear lab coats.

Man-made elements wear robot suits. (Used in Gundam construction.)

スーパー元素周期表

THE SUPER PERIODIC TABLE OF THE ELEMENTS

This is the super periodic table. You can see that the elements get heavier with each row and that the columns are grouped according to their properties. This makes it a very easy-to-understand, illustrative approach to the periodic table.

There is a poster in the back of this book with a larger version of this table, if you'd like to take a closer look.

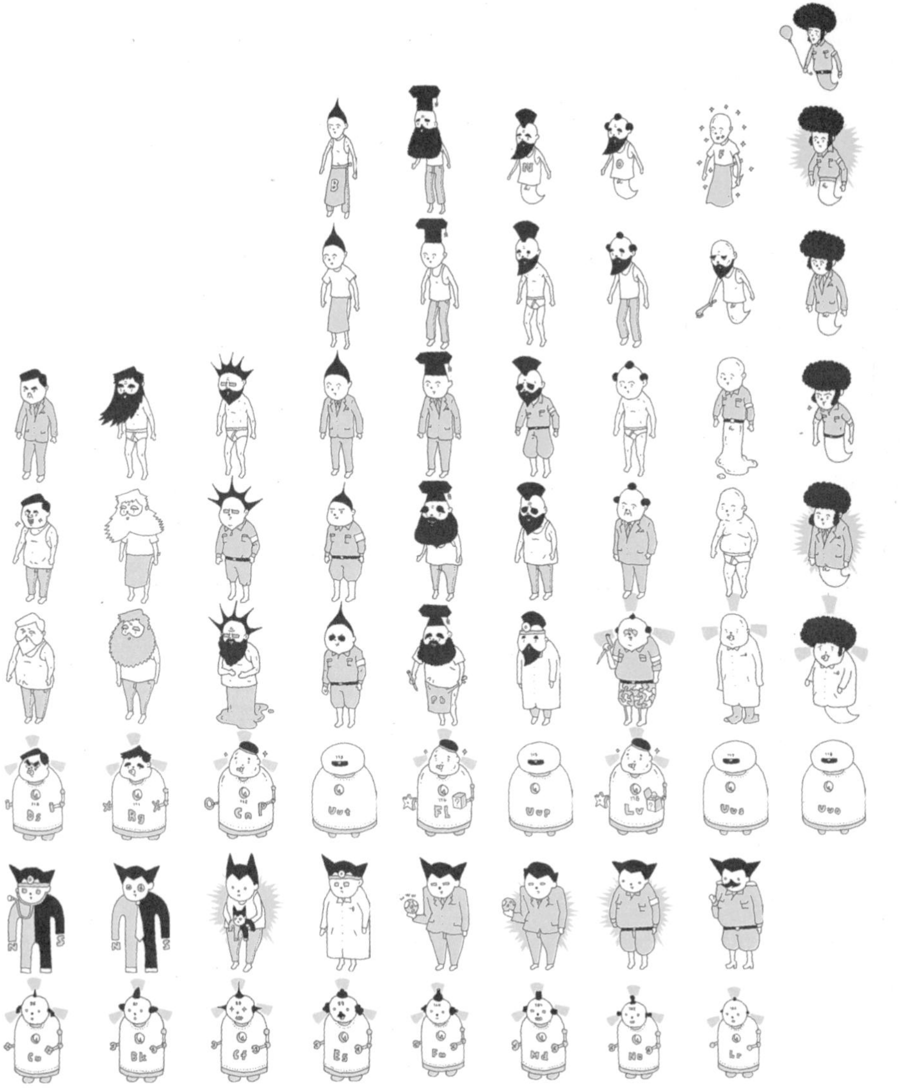
B
N
O
F
Pb
Ds
Rg
Cn
Uut
Fl
Uup
Lv
Uus
Uuo
Cm
Bk
Cf
Es
Fm
Md
No
Lr

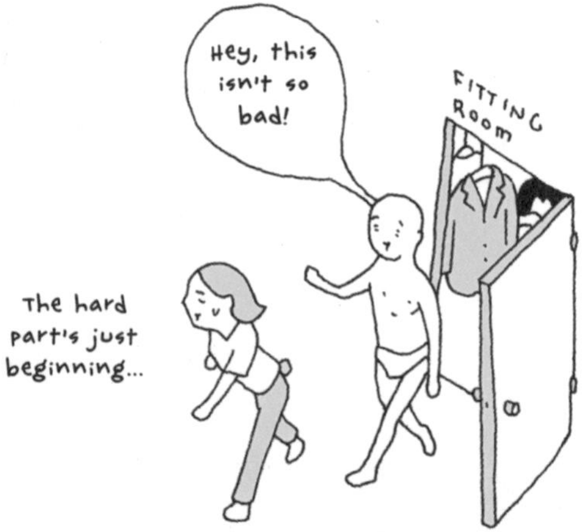
Hey, this isn't so bad!
FITTING ROOM
The hard part's just beginning...

3

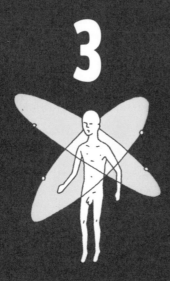

ELEMENT CARTOON CHARACTERS
元素キャラクター

ONE ELEMENT CAN HAVE MANY ROLES.

Now let's take a look at each element individually. What's interesting here is that each element can sometimes be found in the earth, other times in the air, and yet other times inside living beings. Oxygen, for example, erupts in a violent explosion if exposed to fire but turns into water if compounded with hydrogen. Even though we'll be looking at one element at a time, each of them has the potential to fill many different roles. I have therefore tried to limit the information in each presentation to the kind of things that you might encounter in your daily life.

BUT THERE ARE SO MANY OF THEM!

How can a normal human be expected to keep track of them all? Have no fear: If you ever feel lost, just have a look at the following index. The elements are listed in order of atomic number, so finding the one you're looking for should be a piece of cake.

Okay, enough chitchat—on to the elements!

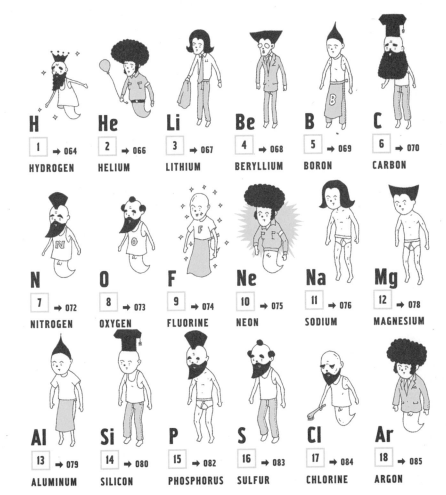

H 1 → 064 HYDROGEN

He 2 → 066 HELIUM

Li 3 → 067 LITHIUM

Be 4 → 068 BERYLLIUM

B 5 → 069 BORON

C 6 → 070 CARBON

N 7 → 072 NITROGEN

O 8 → 073 OXYGEN

F 9 → 074 FLUORINE

Ne 10 → 075 NEON

Na 11 → 076 SODIUM

Mg 12 → 078 MAGNESIUM

Al 13 → 079 ALUMINUM

Si 14 → 080 SILICON

P 15 → 082 PHOSPHORUS

S 16 → 083 SULFUR

Cl 17 → 084 CHLORINE

Ar 18 → 085 ARGON

K
| 19 | → 088 |

POTASSIUM

Ca
| 20 | → 090 |

CALCIUM

Sc
| 21 | → 092 |

SCANDIUM

Ti
| 22 | → 093 |

TITANIUM

V
| 23 | → 094 |

VANADIUM

Cr
| 24 | → 095 |

CHROMIUM

Mn
| 25 | → 096 |

MANGANESE

Fe
| 26 | → 098 |

IRON

Co
| 27 | → 100 |

COBALT

Ni
| 28 | → 101 |

NICKEL

Cu
| 29 | → 102 |

COPPER

Zn
| 30 | → 103 |

ZINC

Ga
| 31 | → 104 |

GALLIUM

Ge
| 32 | → 105 |

GERMANIUM

As
| 33 | → 106 |

ARSENIC

Se
| 34 | → 107 |

SELENIUM

Br
| 35 | → 108 |

BROMINE

Kr
| 36 | → 109 |

KRYPTON

Rb

37 → 112

RUBIDIUM

Sr

38 → 113

STRONTIUM

Y

39 → 114

YTTRIUM

Zr

40 → 115

ZIRCONIUM

Nb

41 → 116

NIOBIUM

Mo

42 → 117

MOLYBDENUM

Tc

43 → 118

TECHNETIUM

Ru

44 → 119

RUTHENIUM

Rh

45 → 120

RHODIUM

Pd

46 → 121

PALLADIUM

Ag

47 → 122

SILVER

Cd

48 → 123

CADMIUM

In

49 → 124

INDIUM

Sn

50 → 125

TIN

Sb

51 → 126

ANTIMONY

Te

52 → 127

TELLURIUM

I

53 → 128

IODINE

Xe

54 → 129

XENON

Cs
| 55 | → 132 |

CESIUM

Ba
| 56 | → 133 |

BARIUM

La
| 57 | → 134 |

LANTHANUM

Ce
| 58 | → 135 |

CERIUM

Pr
| 59 | → 135 |

PRASEODYMIUM

Nd
| 60 | → 136 |

NEODYMIUM

Pm
| 61 | → 137 |

PROMETHIUM

Sm
| 62 | → 137 |

SAMARIUM

Eu
| 63 | → 138 |

EUROPIUM

Gd
| 64 | → 139 |

GADOLINIUM

Tb
| 65 | → 139 |

TERBIUM

Dy
| 66 | → 140 |

DYSPROSIUM

Ho
| 67 | → 140 |

HOLMIUM

Er
| 68 | → 141 |

ERBIUM

Tm
| 69 | → 141 |

THULIUM

Yb
| 70 | → 142 |

YTTERBIUM

Lu
| 71 | → 142 |

LUTETIUM

Hf
| 72 | → 143 |

HAFNIUM

Ta
| 73 | → 143 |

TANTALUM

W
| 74 | → 144 |

TUNGSTEN

Re
| 75 | → 145 |

RHENIUM

Os
| 76 | → 145 |

OSMIUM

Ir
| 77 | → 146 |

IRIDIUM

Pt
| 78 | → 147 |

PLATINUM

Au
| 79 | → 148 |

GOLD

Hg
| 80 | → 149 |

MERCURY

Tl
| 81 | → 150 |

THALLIUM

Pb
| 82 | → 151 |

LEAD

Bi
| 83 | → 152 |

BISMUTH

Po
| 84 | → 152 |

POLONIUM

At
| 85 | → 153 |

ASTATINE

Rn
| 86 | → 153 |

RADON

Fr
87 → 156
FRANCIUM

Ra
88 → 156
RADIUM

Ac
89 → 157
ACTINIUM

Th
90 → 157
THORIUM

Pa
91 → 157
PROTACTINIUM

U
92 → 157
URANIUM

Np
93 → 158
NEPTUNIUM

Pu
94 → 158
PLUTONIUM

Am
95 → 158
AMERICIUM

Cm
96 → 158
CURIUM

Bk
97 → 159
BERKELIUM

Cf
98 → 159
CALIFORNIUM

Es
99 → 159
EINSTEINIUM

Fm
100 → 159
FERMIUM

Md
101 → 160
MENDELEVIUM

No
102 → 160
NOBELIUM

Lr
103 → 160
LAWRENCIUM

Rf
104 → 160
RUTHERFORDIUM

Db
105 → 161
DUBNIUM

Sg
106 → 161
SEABORGIUM

Bh
107 → 161
BOHRIUM

Hs
108 → 161
HASSIUM

Mt
109 → 162
MEITNERIUM

Ds
110 → 162
DARMSTADTIUM

Rg
111 → 162
ROENTGENIUM

Cn
112 → 162
COPERNICIUM

Uut
113 → 163
UNUNTRIUM

Fl
114 → 163
FLEROVIUM

Uup
115 → 163
UNUNPENTIUM

Lv
116 → 163
LIVERMORIUM

Uus
117 → 163
UNUNSEPTIUM

Uuo
118 → 163
UNUNOCTIUM

図の見方

HOW TO READ THE FIGURES

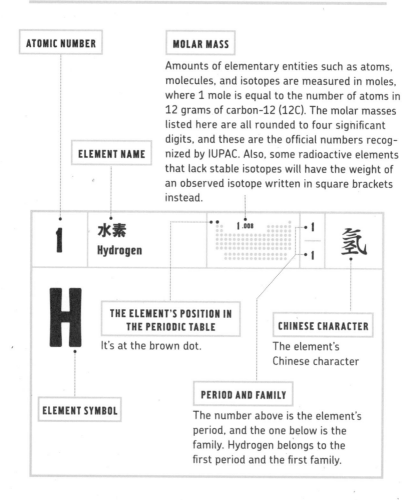

ATOMIC NUMBER

MOLAR MASS

Amounts of elementary entities such as atoms, molecules, and isotopes are measured in moles, where 1 mole is equal to the number of atoms in 12 grams of carbon-12 (12C). The molar masses listed here are all rounded to four significant digits, and these are the official numbers recognized by IUPAC. Also, some radioactive elements that lack stable isotopes will have the weight of an observed isotope written in square brackets instead.

ELEMENT NAME

水素
Hydrogen

1 .008

1

1

氢

THE ELEMENT'S POSITION IN THE PERIODIC TABLE

It's at the brown dot.

CHINESE CHARACTER

The element's Chinese character

PERIOD AND FAMILY

The number above is the element's period, and the one below is the family. Hydrogen belongs to the first period and the first family.

ELEMENT SYMBOL

H

A special element that doesn't fit into any category

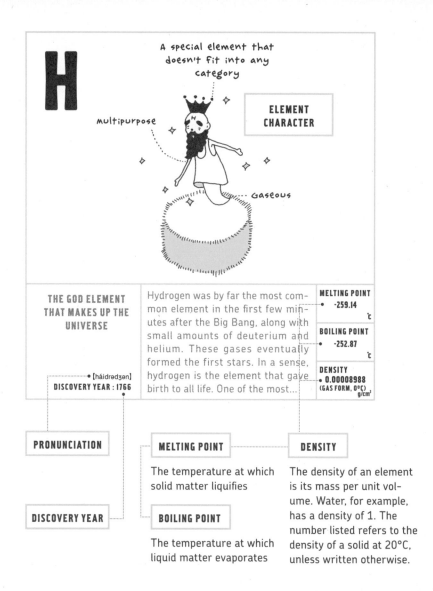

multipurpose

ELEMENT CHARACTER

Gaseous

THE GOD ELEMENT THAT MAKES UP THE UNIVERSE

Hydrogen was by far the most common element in the first few minutes after the Big Bang, along with small amounts of deuterium and helium. These gases eventually formed the first stars. In a sense, hydrogen is the element that gave birth to all life. One of the most...

MELTING POINT
-259.14 ℃

BOILING POINT
-252.87 ℃

DENSITY
0.00008988
(GAS FORM, 0°C)
g/cm³

[háidrədʒən]
DISCOVERY YEAR : 1766

PRONUNCIATION

MELTING POINT

The temperature at which solid matter liquifies

DENSITY

The density of an element is its mass per unit volume. Water, for example, has a density of 1. The number listed refers to the density of a solid at 20°C, unless written otherwise.

DISCOVERY YEAR

BOILING POINT

The temperature at which liquid matter evaporates

周期
PERIOD

1 → 3

原子番号
ATOMIC NUMBER

1→18

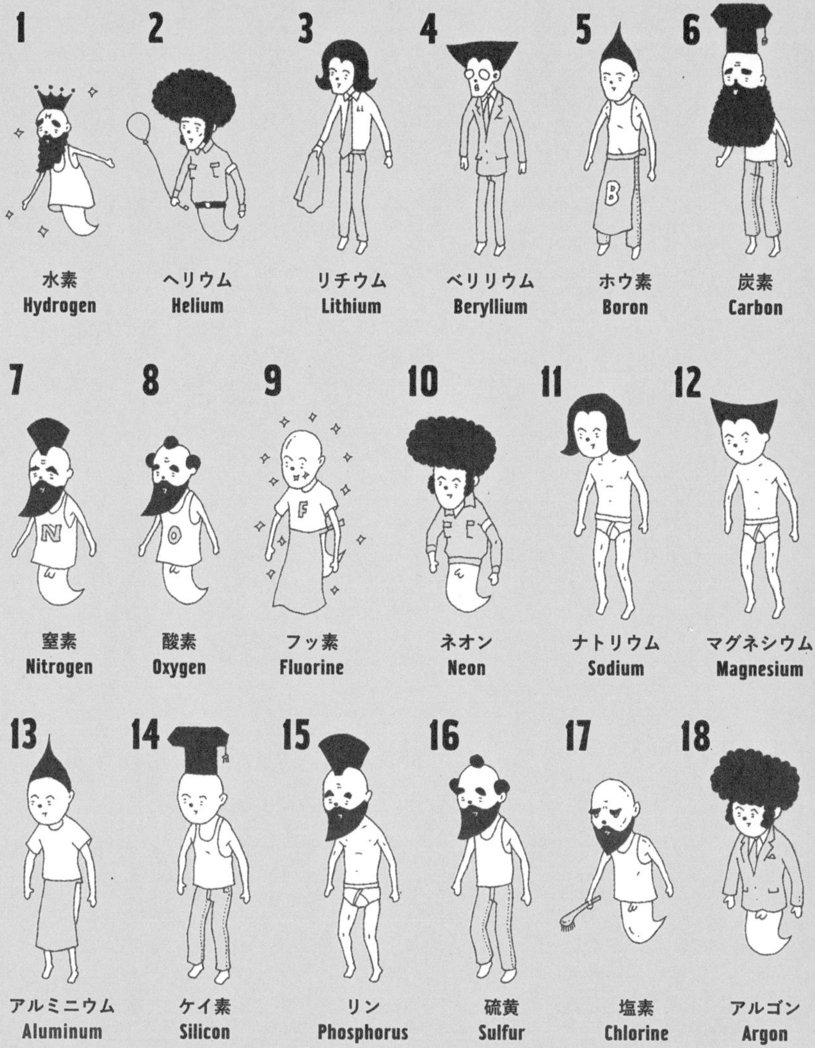
1
水素
Hydrogen
2
ヘリウム
Helium
3
リチウム
Lithium
4
ベリリウム
Beryllium
5
ホウ素
Boron
6
炭素
Carbon
7
窒素
Nitrogen
8
酸素
Oxygen
9
フッ素
Fluorine
10
ネオン
Neon
11
ナトリウム
Sodium
12
マグネシウム
Magnesium
13
アルミニウム
Aluminum
14
ケイ素
Silicon
15
リン
Phosphorus
16
硫黄
Sulfur
17
塩素
Chlorine
18
アルゴン
Argon

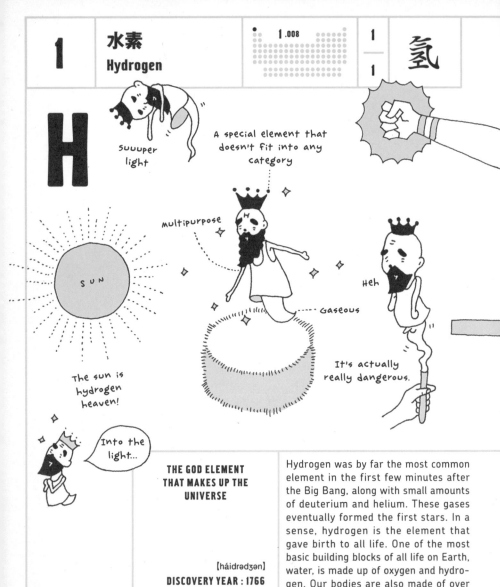

1

水素
Hydrogen

1 .008

1 / 1

氢

H

suuuper light

A special element that doesn't fit into any category

multipurpose

SUN

Gaseous

Heh

It's actually really dangerous.

The sun is hydrogen heaven!

Into the light...

THE GOD ELEMENT THAT MAKES UP THE UNIVERSE

[háidrədʒən]
DISCOVERY YEAR : 1766

Hydrogen was by far the most common element in the first few minutes after the Big Bang, along with small amounts of deuterium and helium. These gases eventually formed the first stars. In a sense, hydrogen is the element that gave birth to all life. One of the most basic building blocks of all life on Earth, water, is made up of oxygen and hydro-gen. Our bodies are also made of over

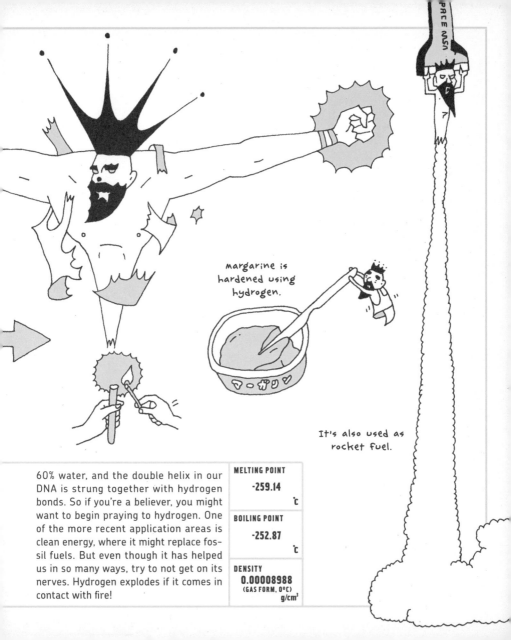

margarine is hardened using hydrogen.

It's also used as rocket fuel.

60% water, and the double helix in our DNA is strung together with hydrogen bonds. So if you're a believer, you might want to begin praying to hydrogen. One of the more recent application areas is clean energy, where it might replace fossil fuels. But even though it has helped us in so many ways, try to not get on its nerves. Hydrogen explodes if it comes in contact with fire!

MELTING POINT
-259.14 ℃

BOILING POINT
-252.87 ℃

DENSITY
0.00008988
(GAS FORM, 0ºC)
g/cm³

2

ヘリウム
Helium

4 .003

1 — 18

氦

He

Found in zeppelins

HELIUM

Noble gas

What? Whoa, it slid out.

very fluid

Gaseous

sound waves

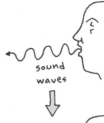

Raises the pitch of your voice

It becomes a wall-climbing liquid at -271°C.

THE LIGHTHEARTED GAS THAT RAISES OUR SPIRITS AND OUR VOICES

[híːliəm]
DISCOVERY YEAR : 1868

Children know it from funny voices and balloons. This ancient element could be found along with hydrogen minutes after the Big Bang. And without these two, no other elements could have been formed. They are the only two elements that are lighter than air, so maybe they're kind of like the leaders, looking down on all the others? But helium, unlike hydrogen, is one cool cookie and doesn't explode easily at all.

MELTING POINT
-272.2
(PRESSURIZED) ℃

BOILING POINT
-268.934 ℃

DENSITY
0.0001785
(GAS FORM, 0°C) g/cm³

3

リチウム
Lithium

6 .941

2 / 1

锂

Li

Battery champion

Lithium ion BATTERY

Li·

GOOD NIGHT

Industrial uses

Alkali metal

Li

solid

Lithium battery

Burns with a bright red color

Beautiful lithium colors

The red in fireworks

THE POWER SOURCE OF THE MOBILE AGE

【líθiəm】
DISCOVERY YEAR : 1817

Lithium, the lightest metal, was also born at the time of the Big Bang, so hydrogen, helium, and lithium are actually triplets. But there was so little lithium at the time, it couldn't do much. Today, however, it is an essential component in both lithium ion batteries and mobile devices. It's light, powerful, and easy to recharge, and it doesn't really deteriorate. It can also be found in seawater, so we won't run out anytime soon.

MELTING POINT
180.54
℃

BOILING POINT
1340
℃

DENSITY
0.534
(0℃)
g/cm³

ベリリウム
Beryllium

9 .012

2 / 2

铍

B e

king of springs

springs that can withstand over 20 billion contractions.

other

Poison

solid

Detrimental to the lungs

POISON

Hard

nya

Light

strong

SUPER TALENTED!
ELITE AND LEGENDARY!

It's the elite metal with skills galore: It weighs two-thirds what aluminum does, it resists heat with a melting point of 1278°C, and it can create springs that can withstand over 20 billion contractions. Yet it still leads a tragic life due to the fact that its particles form a deadly poison. Since it's hard to forge anything without first powdering the materials, it has not been adopted in mass production.

[bəríliəm]
DISCOVERY YEAR: 1797

MELTING POINT
1278±5
℃

BOILING POINT
2970
(PRESSURIZED) ℃

DENSITY
1.8477
g/cm³

5 ホウ素
Boron

10 .81 •

2 / 13

硼

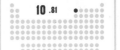

B

Heat-resistant glass

PYREX®

The boron family

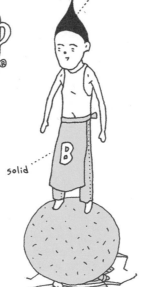

solid

Fake snow in movies

stab

Dehydrates cockroaches using poisonous bait

Disinfecting properties

HELPING OUR DAILY LIVES IN SO MANY WAYS

We mostly use boron in compounds. For example, the technical term for the heat-resistant glass Pyrex is *borosilicate glass*, created by adding boron oxide to keep the glass from swelling and shrinking. Harder diamonds can be created by combining boron with carbon. Finding new boron combinations is a great way for a chemist to show off; two Nobel prizes have been awarded for boron compound research.

[bɔ̀:rɑn]
DISCOVERY YEAR: 1892

MELTING POINT
2300
℃

BOILING POINT
3658
℃

DENSITY
2.34
(TYPE B)
g/cm³

069

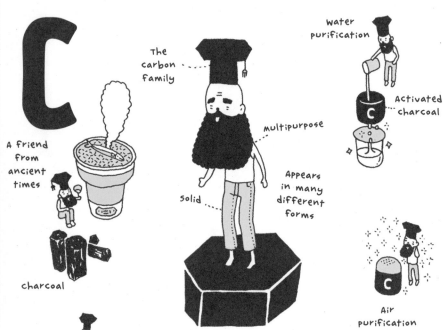

The carbon family

water purification

Activated charcoal

multipurpose

Appears in many different forms

solid

A friend from ancient times

charcoal

Air purification

In calligraphy ink

炭

PART OF EVERY LIVING THING

〔Ká:rbən〕
DISCOVERY YEAR: ANCIENT

Carbon is the building block of all life. One could argue that the food chain should instead be called something like "the carbon tug-of-war." Carbohydrates, proteins, and all the other nutrients that we require are made up of carbon compounds. The same is also true of our cells, DNA, and the plants we eat. (Plants create their carbohydrates from carbon dioxide through a process called

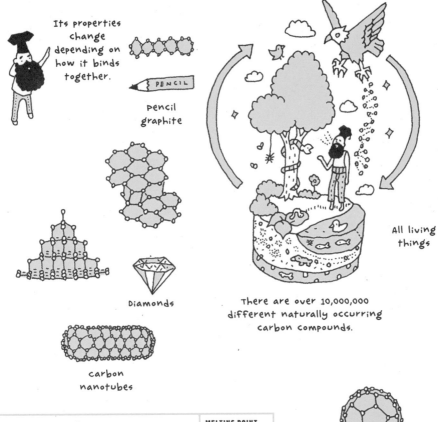

Its properties change depending on how it binds together.

pencil graphite

Diamonds

carbon nanotubes

All living things

There are over 10,000,000 different naturally occurring carbon compounds.

photosynthesis.) The fourth most abundant element in the universe, carbon comes in many forms, from the graphite in our pencils to diamonds. The forms are so different that it's hard to believe that they're made from the same element. It appears today in oil, plastics, clothes, and medicines. It has also drawn a lot of recent attention with the advent of carbon nanotube research.

MELTING POINT
3550
(DIAMOND)
˚c

BOILING POINT
4827
(SUBLIMATION)
˚c

DENSITY
3.513
(DIAMOND)
g/cm³

Fullerene, used in tennis rackets and golfclubs

The nitrogen family

As fertilizer

A dangerous explosive

crash

multipurpose

Gaseous

Liquid nitrogen can chill matter down to -196°C.

TRINITRO TOLUENE
T N T

Air is made up of about 80% nitrogen.

LOOKS FRIENDLY AND COOL, BUT CAN BE DANGEROUS

Making up about 80% of the air we breathe, nitrogen is also the main component of our DNA and the amino acids that make up the proteins in our bodies. It may seem docile, but most explosives—like nitroglycerin and dynamite—are made using nitrogen compounds. Combined with oxygen, it's also a major pollutant. Liquid nitrogen is used in such diverse applications as cryogenics and the preparation of ultra-smooth ice cream.

[náitrədʒən]
DISCOVERY YEAR: 1772

MELTING POINT
-209.86
℃

BOILING POINT
-195.8
℃

DENSITY
0.0012506
(GAS FORM, 0°C)
g/cm³

8

酸素
Oxygen

16 .00 • •

2
16

氧

The oxygen family

made as a by-product from plant photosynthesis

multipurpose

The ozone layer that absorbs harmful ultraviolet radiation

Gaseous

O_3 = ozone

OXYGEN

when something burns, it's actually binding with oxygen.

THE SINGLE-MINDED ELEMENT THAT PROTECTS EARTH

[ɑ́ksidʒən]
DISCOVERY YEAR: 1774

The oxygen most living things need to breathe makes up about 20% of our air and is created primarily through plant photosynthesis. Fire also uses up oxygen when it burns, and the ozone layer that protects us from the sun's ultraviolet rays is made out of it. Rust and rot are also just two types of *oxidation*, which occurs when oxygen binds with different elements and changes their properties.

MELTING POINT
-218.4
℃

BOILING POINT
-182.96
℃

DENSITY
0.001429
(GAS FORM, 0℃)
g/cm³

9	フッ素 Fluorine	19 .00 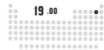	2 / 17	

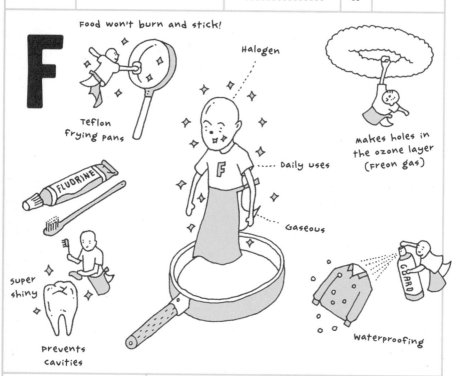

Food won't burn and stick!

Teflon frying pans

Halogen

makes holes in the ozone layer (Freon gas)

FLUORINE

Daily uses

Gaseous

super shiny

prevents cavities

Waterproofing

THE TIDY POISON

When we think of fluorine, we might think of toothpaste or frying pans. It sticks to our teeth after we've brushed them, helping to protect them from bacteria. And coating frying pans and umbrellas with fluorine resin makes it hard for things to stick to them. Pure fluorine, however, is very poisonous, and isolating it from its compounds was no simple feat. The first to do this, the French chemist Moissan, received a Nobel prize.

[flúəriːn]
DISCOVERY YEAR: 1886

MELTING POINT
-219.62 ℃

BOILING POINT
-188.14 ℃

DENSITY
0.001696
(GAS FORM, 0℃)
g/cm³

10	ネオン Neon	20 .18	2 / 18	気

Ne

Noble gas

shiny

NEON SIGN

1912

The first neon sign was made in montmartre, Paris in 1912.

specialist uses

Gaseous

Neon

creates powerful lasers

shines red when subjected to electrical discharge

THE BEACON OF THE NIGHT WAS BORN IN PARIS

[níːan]

DISCOVERY YEAR: 1898

The neon lights that color our cities at night all work by discharging electricity into neon gas encapsulated in glass tubes. The first time this was done was in 1912 in Montmartre, Paris. Neon, normally a very stable gas, shines reddish orange when subjected to electricity. This color can be changed, though, by adding other elements. Helium makes it yellow, mercury makes it turquoise, and argon makes it blue, for example.

MELTING POINT
-248.67 ℃

BOILING POINT
-246.05 ℃

DENSITY
0.00089994
(GAS FORM, 0℃)
g/cm³

| 11 | ナトリウム
Sodium | 22.99 | 3
—
1 | 钠 |

Na

Alkali metal

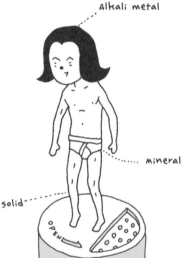

mineral

solid

SALT

Burns yellow

Hydrogen flies away

Reacts explosively with water

makes excellent bathing powder

MOTHER'S FAVORITE, GOOD FOR BOTH FOOD AND CLEANING!

[sóudiəm]
DISCOVERY YEAR: 1807

Sodium compounds are great for housework! For example, table salt (sodium chloride) and baking powder (sodium bicarbonate) are both essential for cooking. Cleaning supplies such as bleaching agents and soaps are based on sodium compounds. Bathing powders and bubble baths are mostly made out of sodium-hydrogen

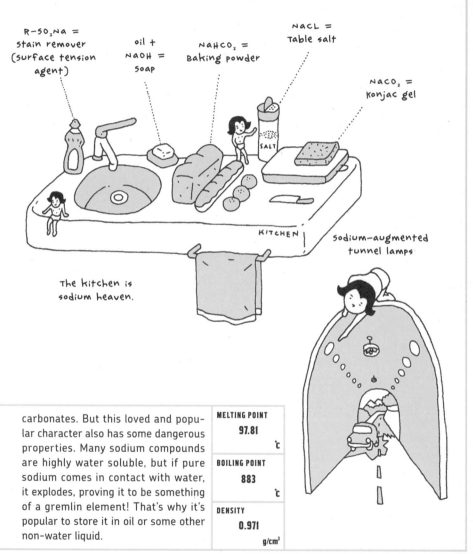

R–SO$_3$Na = stain remover (surface tension agent)

oil + NaOH = soap

NaHCO$_3$ = Baking powder

NaCL = Table salt

NaCO$_3$ = Konjac gel

The kitchen is sodium heaven.

KITCHEN

Sodium-augmented tunnel lamps

carbonates. But this loved and popular character also has some dangerous properties. Many sodium compounds are highly water soluble, but if pure sodium comes in contact with water, it explodes, proving it to be something of a gremlin element! That's why it's popular to store it in oil or some other non-water liquid.

MELTING POINT
97.81 ℃

BOILING POINT
883 ℃

DENSITY
0.971 g/cm³

12

マグネシウム
Magnesium

24.31

3
2

镁

Mg

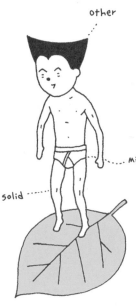

other

solid

mineral

can be found in tofu

Burns brightly

Light but sturdy

Good for making portable electronics

And bricks

THE SUPER SKILLED HONOR STUDENT?!

[mægníːziəm]
DISCOVERY YEAR: 1808

Lighter than aluminum and as strong as steel, magnesium has good electrical and magnetic insulation properties and does not retain heat. That's why it's perfect for laptop and cell phone shells. But magnesium is not just a techie element, as it's found in abundance both in tofu and in the chlorophyll that makes plants green. On top of all these other talents, it's also good for clearing constipation!

MELTING POINT
650
℃

BOILING POINT
1095
℃

DENSITY
1.738
g/cm³

13

アルミニウム
Aluminum

26.98

3 / 13

铝

Al

sapphires

High-voltage wires are made of aluminum.

The boron family

Light

Duralumin cases are made of aluminum alloys.

Daily uses

window frames

solid

can be found in all kinds of everyday things

very conductive

Used in one-yen coins

street signs

Aluminum cans

THE MOST COMMON METAL ON EARTH

[əˈluːmɪnəm]
DISCOVERY YEAR: 1807

Aluminum is a light metal that's very easy to work with. It doesn't rust, conducts electricity well, and is extremely cheap. It can also be alloyed easily to add properties of other metals, producing things like coins, aluminum foil, window frames, and airplane body parts. It has protective properties when applied to stomach membranes and works great as a stress reliever—a good thing in our stress-filled society.

MELTING POINT
660.37 ℃

BOILING POINT
2520 ℃

DENSITY
2.698 g/cm³

| 14 | ケイ素
Silicon | 28.09 • | 3
—
14 | 硅 |

Si

It's sand, basically.

The carbon family

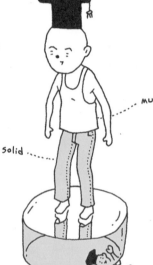

Multipurpose

solid

LSI

The basic material used to make integrated circuits.

For semiconductors

Hee hee

silicone is silicon plastic.

All manner of containers

THE DIGITAL ARTISAN FROM THE DESERT

[sílikən]

DISCOVERY YEAR: 1823

The next time someone asks you about silicon, just point at some sand. It is the second most abundant element on Earth and can be found as silicon dioxide or silicate in (for example) quartz and crystals. In olden times, it was often used for making glass due to its strength, but it's now the mainstay of the digital age. We treasure it as vital to creating semiconductors and solar batteries. Silicone

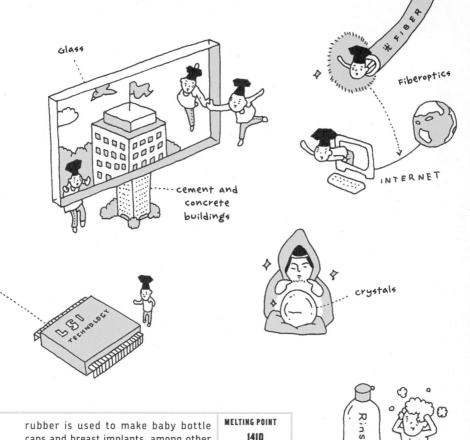

Glass

FIBER

Fiberoptics

cement and
concrete
buildings

INTERNET

LSI TECHNOLOGY

crystals

rubber is used to make baby bottle caps and breast implants, among other things. Silicon dioxide–rich sand has heat-resistant properties and is used to make bricks and building walls. The insulation material asbestos was popular at the end of the 19th century, but now we know that asbestos fibers can accumulate in the lungs and are highly carcinogenic. Pure silicon isn't poisonous at all, though.

MELTING POINT
1410 ℃

BOILING POINT
2355 ℃

DENSITY
2.329 g/cm³

Rinse Si

conditioner

15

リン
Phosphorus

30 .97

3
/
15

磷

P

The nitrogen family

The white parts of bird's nest

mineral

solid

∞

The three essential plant elements

''

N. K. P

Fertilizer

Red phosphorus

Whiff

MARCH

IT ALL STARTED WITH PEE!
THE LIVELY ELEMENT

[fásfərəs]
DISCOVERY YEAR: 1669

About when Isaac Newton was busy dodging falling apples, German alchemists were evaporating urine in their experiments, which led to the discovery of phosphorus. It can be found in several colors, among them white, red, and purple. Our DNA and cells crave it to function properly. It is also essential in agriculture as fertilizer. Red phosphorus is used in the striking surfaces of matches and flares and in cap gun caps.

MELTING POINT
44.2
(WHITE PHOSPHORUS)
℃

BOILING POINT
279.9
(WHITE PHOSPHORUS)
℃

DENSITY
1.82
(WHITE PHOSPHORUS)
g/cm³

硫黄
Sulfur

32.07

3

16

S

The oxygen family

For the skin

Effective as medicine

Antibiotics (penicillin)

can be found in the keratin in our hair

multipurpose

solid

makes onions and garlic spicy

stinky!

Rubber + sulfur =

Tires

THE STINKY VITALITY SOURCE!

[sʌlfər]
DISCOVERY YEAR: ANCIENT

The rotten egg stink of hot springs and the strong smell of garlic and onions are all due to sulfur. But good medicine tastes bitter! The amino acids in our bodies contain sulfur, and sulfur has helped us for decades as part of the world's first antibiotic. Sulfur dioxide, a by-product of combustion engines, is a major pollutant as it can eventually form sulfuric acid in the atmosphere and fall as acid rain.

MELTING POINT
112.8
(CRYSTALLINE FORM)
℃

BOILING POINT
444.674
℃

DENSITY
2.07
(CRYSTALLINE FORM)
g/cm³

17

塩素
Chlorine

35.45

3 / 17

氫

CI

As a swimming pool antibacterial agent

Table salt is a chlorine compound.

SALT

Sodium chloride

Halogen

Multipurpose

Gaseous

chlorine gas is very poisonous.

Detergents

Bleaches

**KILLS BACTERIA!
THE UNRIVALED CLEAN-FREAK**

[klɔ́ːriːn]

DISCOVERY YEAR: 1774

Chlorine is commonly used in water purification plants and pool water as an antibacterial agent. But while it has more or less eradicated epidemic water diseases such as typhoid and cholera, it was also used as a chemical weapon during World War I. It is also used in many everyday items, such as PVC plastics, water pipes, and erasers. Though chlorine itself is very poisonous, chloride ions are necessary to most forms of life.

MELTING POINT
-100.98 ℃

BOILING POINT
-33.97 ℃

DENSITY
0.003214
(0°C)
g/cm³

| 18 | アルゴン
Argon | 39 .95 | 3 / 18 | 氬 |

Ar

Noble gas

Used as a preservative gas

0.93%

N

O

Gaseous

Industrial uses

Used in lightbulbs and fluorescent lamps

The third most common gas in the atmosphere

Insulation glass

AFFABLE AND EASYGOING

Argon gas doesn't react with anything under normal circumstances, which makes it ideal as a preservative for old texts and to isolate experimental materials that react violently with oxygen and hydrogen. It can also be found in fluorescent lights, where it makes it easier for the cathodes in the lamp to discharge electricity. Earth's atmosphere is made up of 78% nitrogen, 21% oxygen, and 1% argon.

[á:rgan]
DISCOVERY YEAR: 1894

MELTING POINT
-189.37
℃

BOILING POINT
-185.86
℃

DENSITY
0.001784
(GAS FORM, 0℃)
g/cm³

周期

PERIOD

4

原子番号

ATOMIC NUMBER

19→36

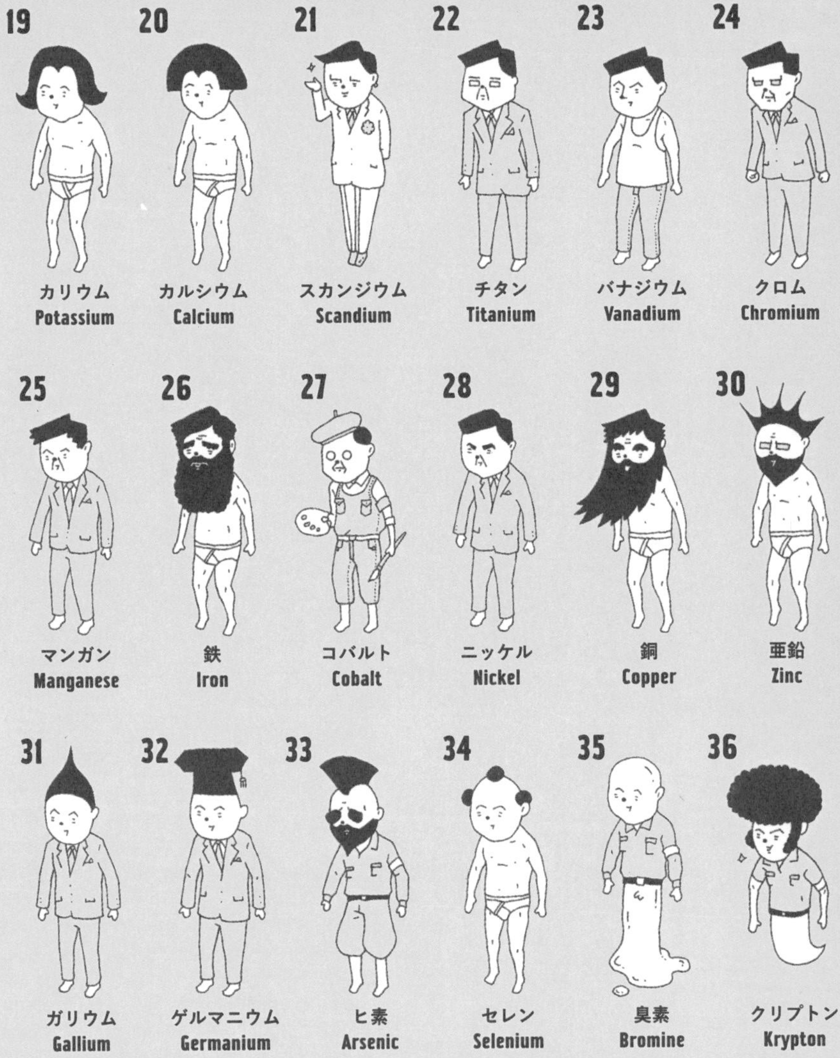
19
カリウム
Potassium
20
カルシウム
Calcium
21
スカンジウム
Scandium
22
チタン
Titanium
23
バナジウム
Vanadium
24
クロム
Chromium
25
マンガン
Manganese
26
鉄
Iron
27
コバルト
Cobalt
28
ニッケル
Nickel
29
銅
Copper
30
亜鉛
Zinc
31
ガリウム
Gallium
32
ゲルマニウム
Germanium
33
ヒ素
Arsenic
34
セレン
Selenium
35
臭素
Bromine
36
クリプトン
Krypton

| 19 | カリウム
Potassium | 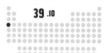 39 .10 | 4
—
1 | 钾 |

K

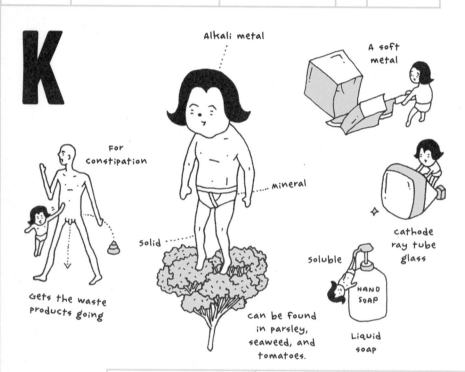

Alkali metal

A soft metal

For constipation

mineral

Solid

Gets the waste products going

cathode ray tube glass

soluble

can be found in parsley, seaweed, and tomatoes.

HAND SOAP

Liquid soap

THE ULTRA-LIVELY MINERAL ELEMENT

[pətǽsiəm]
DISCOVERY YEAR: 1807

Potassium is a mineral that is vital to our bodies and is also one of three main fertilizers used in agriculture. Both potassium and sodium use our cells as their workplace, where they fire nerves and contract muscles. Potassium can also form a multitude of salts with varying properties, depending on which element it bonds with. In addition to the sulfuric and chlorine salts used in fertilizers,

potassium
nitrate in
match heads

5

scrub
scrub

1

2

3

6

Laundry
detergent can
be created by
dissolving plant
potassium in
water.

4

poof

Finely divided potassium
can spontaneously
combust in air, so it's
usually preserved in oil.

potassium fatty acid salts are used in the production of soaps. Potassium nitrate (an ionic salt) is used in fireworks and gunpowder. But even though it's found in many places around the house, potassium is the basis for some very famous poisons. In fact, the poison that we call cyanide is actually a highly soluble compound composed of potassium, carbon, and nitrogen.

MELTING POINT	
63.65	℃

BOILING POINT	
774	℃

DENSITY	
0.862 (-80°C)	g/cm³

20	カルシウム Calcium	40.08	4 / 2	钙

Ca

Alkaline earth metal

Burns with an orange tint

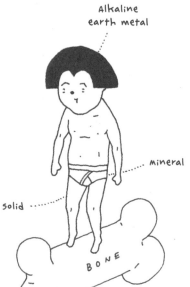

mineral

solid

B O N E

牛乳
MILK

can be found in milk and yogurt

chalk is calcium.

CHALK

tint

**BONES AND SHINING TEETH,
THE WHITE-CLAD WORKER**

[kǽlsiəm]
DISCOVERY YEAR: 1808

Pure calcium is a white metal. It's a well-known ingredient in both yogurt and milk, and it's one of the most sold elements in existence. A grown human body contains approximately 1 kg of calcium, which makes up our skeleton and teeth, among other things. Recent advances in science have enabled us to artificially create the main component of bone, calcium phosphate. This has in turn given us the

marble is also calcium
(calcium carbonate).

Wall plaster is
calcium, too!

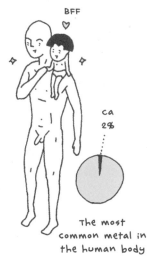

BFF

ca
2%

The most
common metal in
the human body

Limestone
caves

Beautiful

seashells

pearls

Antifreezing
agents used on
roads in winter

technology to manufacture more natural tooth prostheses for people who don't like amalgam fillings. Does it feel kind of strange, knowing that almost all of the minerals in our bodies are actually different kinds of metal? One fun fact is that the nutrients known as vitamins often get discussed together with minerals even though they're not really basic elements. Vitamins are actually organic compounds!

MELTING POINT	
839	°C
BOILING POINT	
1484	°C
DENSITY	
1.55	g/cm³

| 21 | スカンジウム
Scandium | 44 .96 | 4 / 3 | 钪 |

Sc

- Transition metal
- Looks down from above
- Industrial uses
- Solid
- Inside metal halide lamps
- Hisss
- shines incredibly brightly
- STADIUM
- Hah!
- I care not for the masses.
- It's very expensive.

**PRICEY BUT BLAND,
THE SMALL-TIME CELEBRITY**

[skǽndiəm]
DISCOVERY YEAR: 1879

Compared to other elements with a low atomic number, scandium is rare and very expensive. While its weight and other properties are similar to those of aluminum, its melting point is twice as high. A scandium fluorescent tube shines twice as brightly, consumes less electricity, and lasts longer than its halogen counterpart. It's easy to see why these lights are used in high-end cars and stadiums.

MELTING POINT
1541 ℃

BOILING POINT
2831 ℃

DENSITY
2.989 g/cm³

| 22 | チタン **Titanium** | 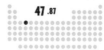 **47**.87 | $\frac{4}{4}$ | 钛 |

Ti

Transition metal

Industrial uses

Dictionary pages

solid

Glasses frames

GOLF HEAD

Dissolves dirt and cleans off water drops

Titanium oxide coating

strong against decay

THE SUPER-USEFUL SMART METAL

[taitéiniəm]
DISCOVERY YEAR: 1795

Used for glasses, piercings, golf clubs, cosmetics, and many other everyday items, titanium was used only for fighter aircraft and submarines until about 30 years ago, when new mining technology brought this metal to the people. It's very nonreactive, able to resist corrosion from both seawater and chemical compounds, and popular among people with metal allergies. It is also light, strong, and abundant.

MELTING POINT
1760
℃

BOILING POINT
3287
℃

DENSITY
4.54
g/cm³

23	バナジウム **Vanadium**	**50**.94 ●	4 — 5	钒

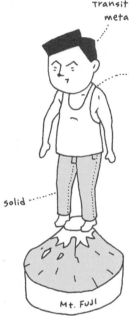

Transition metal

multipurpose

solid

Mt. Fuji

vanadium steel is very hard.

BLUE

Blue paint (vanadium zirconium blue)

Rumored to be good for the health

THE CONTROVERSIAL OCEANIC MINERAL

Some scientists believe that vanadium can have positive effects on your blood sugar levels. Whether this is true or not, the groundwater around Mount Fuji contains lots of it and is therefore sometimes called "Vanadium water." Some types of seaweed and moss are also rich in the mineral, as well as some types of marine invertebrate filter feeders like sea squirts, which have vanadium in their bloodstream.

[vənéidiəm]
DISCOVERY YEAR: 1830

MELTING POINT
1887
℃

BOILING POINT
3377
℃

DENSITY
6.11
(19℃)
g/cm³

24

クロム
Chromium

52 .00

4 / 6

铬

Cr

Transition metal

Industrial uses

chrome yellow is very beautiful.

Solid

chrome plating

Doesn't rust

STAINLESS

stainless steel
Fe — Ni — Cr

Brings out the color in emeralds and rubies

Hexavalent chromium is very toxic.

CHROME YELLOW

THE TORTURED ARTIST

[króumiəm]
DISCOVERY YEAR: 1797

Many have lost trust in chromium because of pollution issues. But these stem mainly from the hexavalent chromium oxidation state, while the trivalent state is an essential trace mineral. Chromium is also the basis for many beloved hues, such as viridian and the vivid colors of emeralds and rubies. And it is one of the components of stainless steel. One hopes that its accomplishments have garnered it a little honor.

MELTING POINT
1857 ℃

BOILING POINT
2672 ℃

DENSITY
7.19 g/cm³

マンガン
Manganese

54.94

4/7

錳

Mn

Fe + Mn = a very strong compound

Transition metal

Industrial uses

Solid

The cables on the great seto bridge are made of manganese steel.

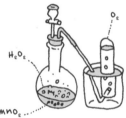

H₂O₂

O₂

MnO₂

Manganese dioxide

often used in scientific experiments

A WORKER OF OLD, THE UNSUNG HERO OF THE ELEMENTS

[mǽŋgǝníːs]
DISCOVERY YEAR: 1774

Famous as the raw material for dry cell batteries, manganese is a metal found both on dry land and on the sea floor. But while manganese batteries have been in use since the late 19th century, they are gradually being replaced by the alkali family of batteries (though actually there isn't much difference between the materials used in these two battery types). Manganese is also necessary for our metabolism.

MELTING POINT	1244 ℃
BOILING POINT	1962 ℃
DENSITY	7.44 g/cm³

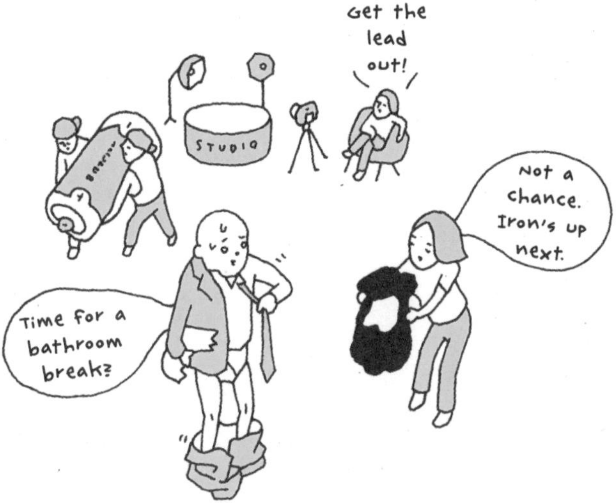
Get the lead out!
STUDIO
Not a chance. Iron's up next.
Time for a bathroom break?

26

鉄
Iron

55.85

4 / 8

铁

Fe

Transition metal

Trains

Ships

mineral

cars

Body warmers

solid

Tapes

THE COGWHEEL OF DESTINY THAT SET CIVILIZATION IN MOTION

〔áiərn〕
DISCOVERY YEAR: ANCIENT

The discovery of iron was the turning point for all humankind, allowing us to throw away our stone tools and set out on the path to civilization. The first people to use iron were the ancient Hittites in 1500 BCE. After their kingdom fell, the Hittite people spread across the globe, taking their craft with them and bringing a gradual but significant change to people's lives. Iron still accounts for

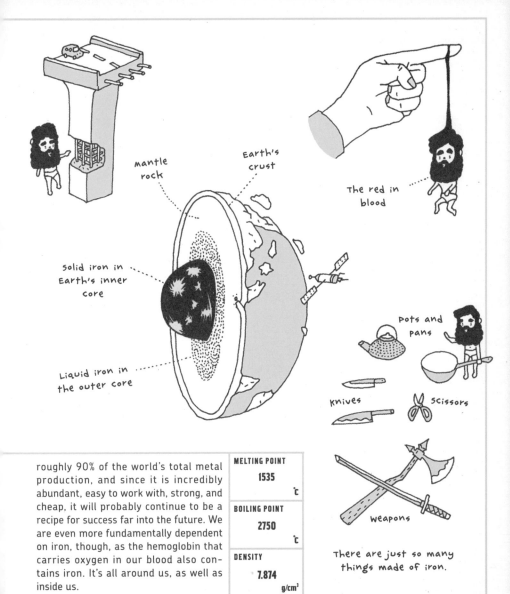

mantle rock

Earth's crust

The red in blood

Solid iron in Earth's inner core

Liquid iron in the outer core

Pots and pans

knives

scissors

weapons

There are just so many things made of iron.

roughly 90% of the world's total metal production, and since it is incredibly abundant, easy to work with, strong, and cheap, it will probably continue to be a recipe for success far into the future. We are even more fundamentally dependent on iron, though, as the hemoglobin that carries oxygen in our blood also contains iron. It's all around us, as well as inside us.

MELTING POINT
1535 ℃

BOILING POINT
2750 ℃

DENSITY
7.874 g/cm³

コバルト
Cobalt

58 .93

4
9

钴

Co

eye wash

Eye drops

magnets

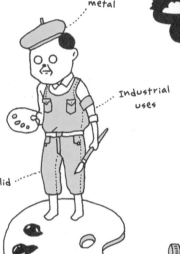

Transition metal

Industrial uses

Solid

No art without cobalt

cobalt BLUE

cobalt GREEN

THE BLUE-CLAD DIGITAL TECHNICIAN

[kóubɔːlt]

DISCOVERY YEAR: 1737

You probably know cobalt from its charming signature color, cobalt blue, but did you know that its name comes from the German word *kobold*, which means goblin? Silver miners in 18th century Germany simply didn't know how to react when they encountered veins of this ghastly blue metal that gave off toxic fumes. Nowadays its magnetic and sensitive properties make it ideal for use in computer hard disks and many other items.

MELTING POINT

1495
℃

BOILING POINT

2870
℃

DENSITY

8.9
g/cm³

28

ニッケル
Nickel

58.69

4 / 10

镍

Ni

Transition metal

Industrial uses

solid

underwire bras

charging nickel-metal hydride batteries with solar panels

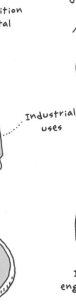

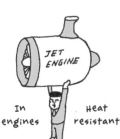

JET ENGINE

In engines

Heat resistant

THE MONEY MAKER

[nÍkəl]
DISCOVERY YEAR: 1751

The copper-nickel alloy *cupronickel* is used in American nickels and in Japanese 100-yen and 50-yen coins. Over 1,000,000 tons of nickel are produced worldwide every year. The metal is used in a multitude of alloys, especially iron alloys like stainless steel but also shape-memory titanium alloys. Nickel has gotten a lot of attention lately with the advent of environmentally friendly nickel-metal hydride rechargeable cell batteries.

MELTING POINT
1455
℃

BOILING POINT
2890
℃

DENSITY
8.902
(25℃)
g/cm³

| 29 | 銅
Copper | 63.55
•
4
11 | 铜 |

copper statues

Cu

Transition metal

solid

mineral

spider, snail, and octopus blood

copper wires are very conductive.

10-yen coins are made of bronze.

THE METAL WE'VE CARED FOR THE LONGEST

[kápər]
DISCOVERY YEAR: ANCIENT

The oldest known man-made metal object is a 10,000-year-old copper pendant found in Iraq. Copper conducts heat well and is easy to work with. It's too brittle to use for anything other than household tools, but alloying copper with tin to produce bronze made it possible to construct weapons, musical instruments, farming tools, and more—an event so important, we call it the Bronze Age. Copper deserves a gold medal!

MELTING POINT
1083.5 ℃

BOILING POINT
2567 ℃

DENSITY
8.96 g/cm³

| 30 | 亜鉛
Zinc | 65.38
● | 4
12 | 锌 |

Zn

Fe + zn plating

Galvanized sheet metal is great for water buckets and roofs.

The zinc family

mineral

solid

Tasty.

oysters contain a lot of zinc.

copper + zinc = brass

THE PICKY GOURMET ELEMENT

[zínk]
DISCOVERY YEAR: MEDIEVAL

Zinc is a very important trace mineral, second in our bodies only to iron. For example, it helps the tongue cells in our taste buds process our sense of taste. This is why zinc deficiencies often lead to an impaired appetite. It's also an excellent construction material, creating alloys such as galvanized sheet metal with iron and brass with copper. It has also recently been used as raw material in creating blue LEDs.

MELTING POINT
419.58
℃

BOILING POINT
907
℃

DENSITY
7.133
g/cm³

ガリウム
Gallium

69.72

4 / 13

镓

Ga

The boron family

For reading compact discs

Industrial uses

solid

Green light diodes

Electronic gizmos

THE KIND, NERDY ELEMENT

[gǽliəm]
DISCOVERY YEAR: 1875

Are you wondering, "WTF is gallium?" Well, you should be ashamed! In addition to being a vital part of both game consoles and Blu-ray players, it's also used in semiconductors and LEDs. Gallium nitride is in almost all new video equipment, driving the powerful blue lasers that were unattainable with lesser technology. This has allowed us to achieve higher resolutions, sharper colors, and a more awesome entertainment experience.

MELTING POINT
29.78
℃

BOILING POINT
2403
℃

DENSITY
5.907
g/cm³

32	ゲルマニウム **Germanium**	⋯ **72**.64 ⋯	$\dfrac{4}{14}$	锗

Ge

The carbon family

Industrial uses

solid

TRAN-SISTOR

Nostalgia...

The first germanium radio

In wide-angle camera lenses

Not very popular nowadays

THE ELEMENT FROM THE GOOD OLD DAYS

[dʒərméiniəm]
DISCOVERY YEAR: 1885

This element might be familiar to the audiophiles out there, since the heart of the world's first transistor radio (produced by Sony in 1953) was made of germanium. It was used widely at the dawn of the semiconductor age but has since been replaced by other elements. Recent rumors hint that it might be good for the health, though, with its name appearing on several products such as "germanium hot baths."

MELTING POINT
937.4 ℃

BOILING POINT
2830 ℃

DENSITY
5.323 g/cm³

| 33 | 匕素
Arsenic | 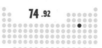 74 .92 | 4
—
15 | 砷 |

As

Used in semiconductors with gallium and indium

can be found in some types of edible seaweed

The nitrogen family

specialist uses

solid

D'oh.

Used as a poison too many times

Found in our bodies!

It can also be used to make medicine.

THE RUTHLESS DARK-SIDE ELEMENT

[ά:rsənik]
DISCOVERY YEAR: MEDIEVAL

Most people probably know arsenic as a poison, rumored to be responsible for the deaths of Napoleon Bonaparte and King George III. It blocks enzymes when introduced to the bloodstream, and it is both odorless and tasteless, which makes it very hard to detect when hidden in food. Some types of seaweed naturally contain arsenic, but not enough to make you sick. Arsenic is widely used for making semiconductors.

MELTING POINT
817
(METAL, PRESSURIZED) ℃

BOILING POINT
616
(SUBLIMATION) ℃

DENSITY
5.78
(METAL) g/cm³

34	セレン **Selenium**	78.96	4	硒
			16	

Se

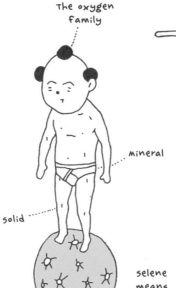

The oxygen family

mineral

solid

selene means "moon" in Greek.

Used to make windows for skyscrapers

It is important to our bodies.

Japan produces the most selenium in the world!

**GOOD AND EVIL,
THE ELEMENT WITH TWO FACES**

[síli:niəm]
DISCOVERY YEAR: 1817

Selenium is pretty smelly, as it belongs to the same family as sulfur, but it's a vital part of our metabolism. A selenium deficiency makes your immune system weaker, but if you take too much, it can damage your intestines and stomach! It's pretty easy to take in just the right amount, as shellfish, vegetables, beef, eggs, and many other foods contain selenium in small quantities. Selenium is also used in night-vision cameras.

MELTING POINT
217
℃

BOILING POINT
684.9
℃

DENSITY
4.79
(GRAY SOLID)
g/cm³

臭素
Bromine

79 .90

4
—
17

溴

Br

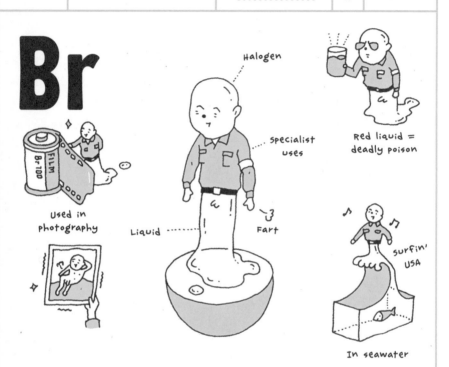

Halogen

Specialist uses

Red liquid = deadly poison

Used in photography

Liquid

Fart

surfin' USA

In seawater

MORE ROMANTIC THAN IT SOUNDS

[bróumi:n]
DISCOVERY YEAR: 1826

The French chemist Antoine Jérôme Balard and the German chemist Carl Jacob Löwig each independently discovered bromine as students in 1826. Bromine dyes (extracted from certain species of snails) were sought after in ancient Japan and Europe for their beautiful color, a vivid purple. Silver bromide is also very sensitive to light, which has made it the basis of modern photography materials.

MELTING POINT
-7.3
℃

BOILING POINT
58.78
℃

DENSITY
3.1226
(LIQUID, 20°C)
g/cm³

クリプトン
Krypton

83 .80

4 / 18

氪

Kr

Noble gas

very rare

specialist uses

Gaseous

The name of superman's home planet

Bright

Krypton light bulbs

THE BRIGHTLY SHINING FLASH-MAN

[kríptan]
DISCOVERY YEAR: 1898

Most people probably know that Superman's home planet is named Krypton, but the element's name actually comes from the word *cryptic*, as it was very hard to discover. Krypton light bulbs can be made very small and still outshine any argon-based counterpart, which makes them popular with photographers and filmmakers. Krypton is also used in stroboscopes, high-powered gas lasers, and many other applications.

MELTING POINT
-156.6
°c

BOILING POINT
-152.3
°c

DENSITY
0.0037493
(GAS FORM, 20°C)
g/cm³

周期
PERIOD

5

原子番号
ATOMIC NUMBER

37→54

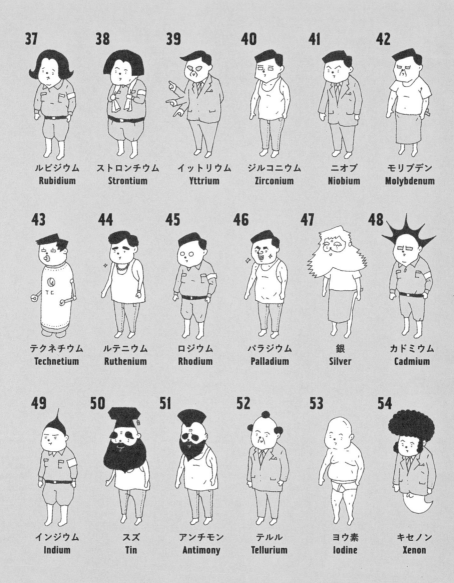

37 ルビジウム Rubidium

38 ストロンチウム Strontium

39 イットリウム Yttrium

40 ジルコニウム Zirconium

41 ニオブ Niobium

42 モリブデン Molybdenum

43 テクネチウム Technetium

44 ルテニウム Ruthenium

45 ロジウム Rhodium

46 パラジウム Palladium

47 銀 Silver

48 カドミウム Cadmium

49 インジウム Indium

50 スズ Tin

51 アンチモン Antimony

52 テルル Tellurium

53 ヨウ素 Iodine

54 キセノン Xenon

| 37 | ルビジウム
Rubidium | 85.47 | 5
—
1 | 鉫 |

Rb

Used to measure the age of rocks

Used in cathode ray tube glass

Alkali metal

Specialist uses

solid

clock

Explodes violently if it touches water

Atomic clocks made with rubidium have a yearly error of 0.1 seconds.

THE TIMEKEEPER OF THE UNIVERSE

Tick tock. The atomic clock that controls the NHK time broadcasts* works by monitoring the energy fluctuations of a rubidium isotope and misses by only 1 second every 10 years or so. The half-life of rubidium is a whopping 48.8 billion years, perfect for assessing the age of Earth's minerals and asteroid remnants. This is done by measuring the rubidium left in the sample, then calculating how long it took to decay to that point.

[ru:bídiəm]
DISCOVERY YEAR: 1861

MELTING POINT
39.1
℃

BOILING POINT
688
℃

DENSITY
1.532
g/cm³

* NHK shows the current time in the corner of all its TV broadcasts.

Sr

Alkaline earth metal

Red

In flares

specialist uses

solid

Fireworks

Pop

The vivid red in fireworks

THE SWEET FIREBALL DUDE

[stránʃiəm]
DISCOVERY YEAR: 1787

The scarlet explosions that stand out in any fireworks show are probably made of strontium. All alkali and alkaline earth metal elements burn with different colors, but strontium outshines the rest with its brilliant hue. It's also used in most commercial flares. It takes after its alkaline earth metal big brother, calcium, in that it is easily absorbed into bone. This is why it's also used for bone tumor treatments and diagnostic measures.

MELTING POINT
769
℃

BOILING POINT
1384
℃

DENSITY
2.54
g/cm³

| 39 | イットリウム
Yttrium | 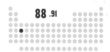 88 .91 | 5
—
3 | 钇 |

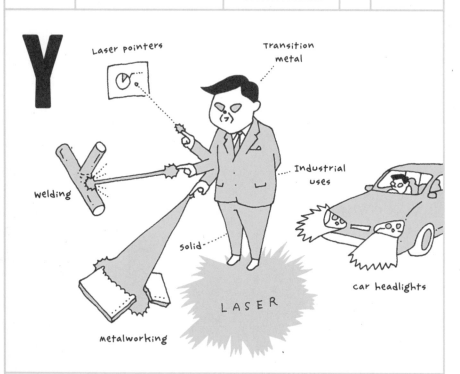

Laser pointers

Transition metal

welding

Industrial uses

solid

car headlights

L A S E R

metalworking

THE PIONEER OF THE LASER WORLD

[ítriəm]
DISCOVERY YEAR: 1794

I'm guessing most of us played with pocket lasers as kids, but did you know that *laser* is an acronym that stands for "Light Amplification by Stimulated Emission of Radiation"? A mouthful, huh? Yttrium and aluminum oxides are used in the creation of YAG crystals, which are vital to the construction of solid-state lasers. They're used in factories and hospitals as welding and operating-room tools.

MELTING POINT
1522
℃

BOILING POINT
3338
℃

DENSITY
4.469
g/cm³

| **40** | ジルコニウム
Zirconium | **91**.22 | 5
—
4 | 锆 |

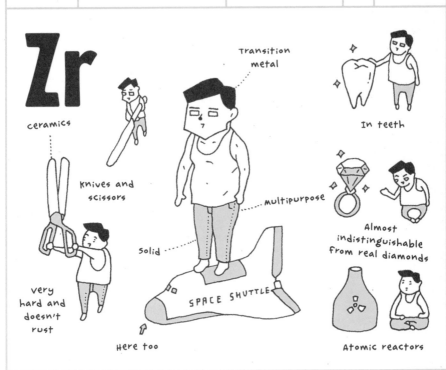

Zr

ceramics

Transition metal

In teeth

knives and scissors

multipurpose

Almost indistinguishable from real diamonds

Solid

very hard and doesn't rust

SPACE SHUTTLE

Here too

Atomic reactors

DIAMONDS FOR EVERYONE!

[zəːrkóuniəm]
DISCOVERY YEAR: 1789

Zirconium shines as brightly as any diamond if processed correctly (as cubic zirconia). It can also be made into a rust-free ceramic material that's harder than steel if it's oxidized, ground into a powder, and sintered. These advanced ceramics can be used for creating useful household tools such as scissors and kitchen knives, as well as in more exotic applications like spacecraft and jet engines.

MELTING POINT
1852
℃

BOILING POINT
4377
℃

DENSITY
6.506
g/cm³

115

ニオブ
Niobium

92.91

5 / 5

铌

Nb

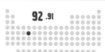
Transition metal

zero resistance

Iron Niobium
strong

Ferroniobium

used in pipelines

Industrial uses

superconductivity occurs at low temperatures.

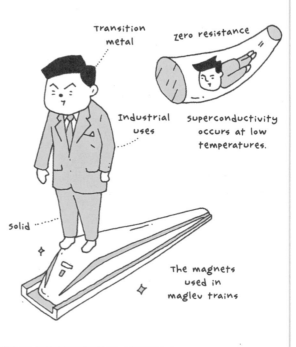
Solid

The magnets used in maglev trains

SUPPORTING THE PRACTICALITIES OF THE FUTURE

[naióubiəm]
DISCOVERY YEAR: 1801

Niobium is named after Niobe, the daughter of Tantalus in Greek myth, since it bears some resemblance to element 73 (tantalum). But despite the name's ancient origins, it now represents an element used in cutting-edge jet engines, space shuttles, and maglev vehicles. The metal can create extremely powerful magnetic materials by being alloyed with steel. This makes it not only heat resistant but also superconductive.

MELTING POINT
2468 ℃

BOILING POINT
4742 ℃

DENSITY
8.57 g/cm³

Mo

Transition metal

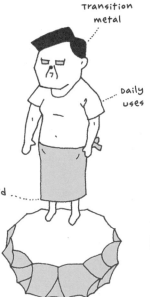

Daily uses

Solid

pfft

—3

ultra-modern toilet seats

The strongest steel is made with molybdenum.

Industrial lubricants

THE DIVERSE BLACKSMITH

[məlíbdənəm]
DISCOVERY YEAR: 1778

Molybdenum steel is a very strong and rust-resistant iron alloy. Knives made from this steel can cost several hundred dollars. This specialist material is also used in jet plane landing gear and rocket engines. Recent research has enabled us to use molybdenum to heat water more effectively, creating a new generation of ceramic heaters (used in automated Japanese toilets, which use warm jets of water instead of toilet paper).

MELTING POINT
2617
℃

BOILING POINT
4612
℃

DENSITY
10.22
g/cm³

43	テクネチウム **Technetium**	[99] ●	5 — 7	锝

Tc

Transition metal

man-made

medical

Tc

Used to find blood clots

solid

continuously falls apart

**THE FIRST
MAN-MADE ELEMENT**

While there might have been particles of the 43rd element at the time Earth was born, they have long since decayed. Scientists searched for this element for decades after Mendeleev predicted its existence. The element has many medical uses. For example, because the technetium-99m isotope decays very quickly, it is used as a radioactive tracer to perform imaging scans and detect blood clots.

[tekníːʃiəm]
DISCOVERY YEAR: 1936

MELTING POINT
2172
℃

BOILING POINT
4877
℃

DENSITY
11.5
g/cm³

| 44 | ルテニウム
Ruthenium | 101.1
• | 5
—
8 | 钌 |

Ru

Transition metal

Multipurpose

Hard but brittle

Fountain pen tips

ruthenium

Solid

Good at splitting water into hydrogen and oxygen with the help of sunlight

Light

Increasing the size of hard disks

Hard disk

A CELEBRITY SINCE BIRTH

While it hangs out with the other precious metals, ruthenium isn't really an accessory type of guy. However, it did contribute to two recent Nobel prizes (in 2001 and 2005) as a catalyst in organic synthetic chemistry. It's great for creating higher-capacity magnetic hard drives, and since it has a beautiful luster and is durable, it's also used for making fountain pens. An air of glamour hangs about this element.

[ruːθíːniəm]
DISCOVERY YEAR: 1844

MELTING POINT
2310
℃

BOILING POINT
3900
℃

DENSITY
12.37
g/cm³

| 45 | ロジウム
Rhodium | 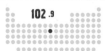 102.9 | 5 / 9 | 銠 |

Rh

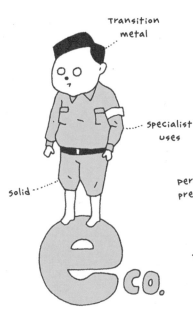

Transition metal

specialist uses

solid

e co.

It's actually very rare.

perfectly preserved

For polishing jewelry

As a purifier catalyst

Rhodium

ALWAYS A BRIDESMAID, NEVER A BRIDE

Only 16 tons of this precious metal are produced every year. And even though it's of higher quality than both gold and platinum, it's never allowed up on the main stage. However, it does participate—it's used as a coating material. Its beautiful white color doesn't lose its shine over time, and it makes silver and platinum last longer when processed together. This admirable element supports others at the cost of its own fame.

[róudiəm]
DISCOVERY YEAR: 1803

MELTING POINT	
1966	℃

BOILING POINT	
3727	℃

DENSITY	
12.41	g/cm³

| 46 | パラジウム
Palladium | 106.4
• | 5
—
10 | 钯 |

Pd

can hold 900 times its own volume in hydrogen

Pd H BANK

Transition metal

multipurpose

solid

crack

palladium and gold/silver alloys

THE FORMER UGLY DUCKLING

[pəléidiəm]
DISCOVERY YEAR: 1803

Long long ago, it was considered bad luck when a gold miner found a vein contaminated by palladium. The element was found about the same time that the asteroid Pallas was discovered and was therefore named after it. It is well liked by scientists as it can hold up to 900 times its own volume in hydrogen. Palladium is used in hydrogen fuel cells and as a catalyst when producing organic compounds. It's also used in dentistry.

MELTING POINT
1552 ℃

BOILING POINT
3140 ℃

DENSITY
12.02 g/cm³

121

| 47 | 銀
Silver | 107.9 | 5
—
11 | 银 |

As accessories and utensils

Transition metal

Daily uses

Solid

Silver nitrate compounds are used in photo paper.

For warding off demons

STYLISH AND GOOD AT WHAT HE DOES

[silver]
DISCOVERY YEAR: ANCIENT

Silver's shine evokes a romantic mood, and this metal is cheap and easy to work with. This makes it perfect for utensils and accessories. Silver ions are also particularly good at killing bacteria by destabilizing their enzymes, and silver's gaining ground as a component of deodorants and odor-resistant fibers. Its natural enemy is sulfur, which on contact makes silver go black. So don't wash your silverware in the local hot spring!

| MELTING POINT |
| 961.93 ℃ |

| BOILING POINT |
| 2212 ℃ |

| DENSITY |
| 10.5 g/cm³ |

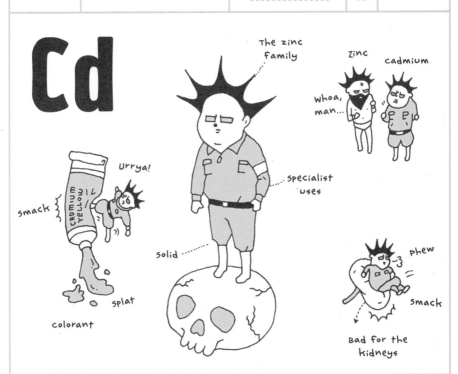

The zinc family

zinc cadmium

Whoa, man...

Urrya!

smack

CADMIUM YELLOW

Specialist uses

solid

splat

colorant

phew

smack

Bad for the kidneys

THE RAMPAGING MAD SCIENTIST

[kǽdmiəm]
DISCOVERY YEAR: 1817

A mysterious sickness that spread near the Jinzuu River from 1912 to 1946 became known as one of the four big pollution diseases of Japan and was called the *itai-itai* ("ouch-ouch") disease. It was caused by cadmium from a mine upstream. Since it's very similar in structure to zinc, cadmium can enter the body, where it eventually weakens bones and obstructs the kidneys. Uses include pigments and nickel-cadmium batteries.

MELTING POINT
320.9
℃

BOILING POINT
765
℃

DENSITY
0.00865
(25℃)
g/cm³

49

インジウム
Indium

114.8

5 / 13

铟

In

sprinkler systems

Dissolves well in heat

The boron family

commonly recycled

specialist uses

solid

LCDs

most of it is produced in china.

HE'S IN SEASON!
THE HERO OF THE DAY

[índiəm]
DISCOVERY YEAR: 1863

Indium is indispensable to electronics manufacturers, as it's used for making flat-screen TVs. Its unusual quality of being able to create transparent and conductive films is vital for making all types of LCD, plasma, and OLED* displays. Japan was once the world's largest producer of indium, but since the mine shut down in 2006, people are now scrambling to enact indium recycling programs all over the world.

MELTING POINT
156.17
℃

BOILING POINT
2080
℃

DENSITY
7.31
(25°C)
g/cm³

* OLED is a type of light-emitting diode made up of organic compounds.

| 50 | スズ
Tin | ⬤ 118.7 | 5 / 14 | 錫 |

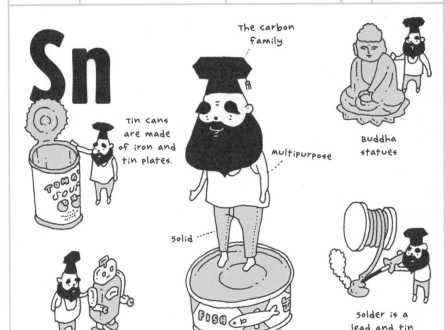

Sn

The carbon family

Tin cans are made of iron and tin plates.

multipurpose

solid

Buddha statues

solder is a lead and tin alloy.

Toys

THE HERO OF OLD TURNED SLACKER

[tín]
DISCOVERY YEAR: ANCIENT

Tin is abundant, easy to work with, and has a low melting point. Its alloy with copper, bronze, has been used throughout history to make swords and spear tips. It has also been used in Japan since the Nara period for building Buddha statues. Despite having been used to make almost everything, it has few uses left. It can still be found in tin model toys, tin cans, solder, and printing equipment, though.

MELTING POINT
231.9681
℃

BOILING POINT
2270
℃

DENSITY
7.31
(WHITE TIN)
g/cm³

51 アンチモン
Antimony

121.8

5 / 15

锑

Sb

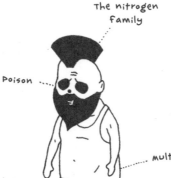

The nitrogen family

Poison

Used in linotype machines

Multipurpose

Solid

cleopatra's makeup

poisonous

ANTI

makes things harder to burn

CLEOPATRA'S DARLING

You don't see it often these days, but antimony is used in some semiconductors and in the poles of lead batteries. It's also used together with lead in printing equipment and is steadily gaining ground in other areas. In ancient Egypt antimony sulfide (as kohl) was Queen Cleopatra's eyeliner of choice—a pretty glamorous past for such a steady worker. I wouldn't recommend using it the same way now, though, as it's rather toxic.

【ǽntəmòuni】
DISCOVERY YEAR: 1450

MELTING POINT
630.74
℃

BOILING POINT
1635
℃

DENSITY
6.691
g/cm³

| 52 | テルル
Tellurium | 127.6 | 5 / 16 | 碲 |

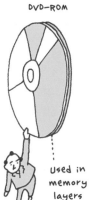

DVD-ROM

Used in memory layers

The oxygen family

Industrial uses

solid

DVD

Wine cellars

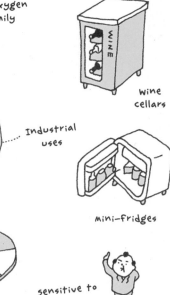

mini-fridges

sensitive to temperature

THE CUTE BUT SMELLY ELEMENT

[telúəriəm]
DISCOVERY YEAR: 1782

The quaint-sounding element tellurium is named after the Latin word for our planet, *Tellus*, and is used in everything from DVD data recording to green LEDs. It's also great for making quiet and versatile mini-fridges when compounded with bismuth and selenium. It can be alloyed with iron, copper, and lead to make these metals easier to work with. It's too bad that it smells like garlic, which makes it a bit hard to be around.

MELTING POINT
449.5 ℃

BOILING POINT
990 ℃

DENSITY
6.24 g/cm³

| 53 | ヨウ素
Iodine | 126.9

•
5 / 17 | 碘 |

Iodine mouthwash

インジン®
ガーヴル

IODINE

Disinfectant

Halogen

solid.

mineral

Japan is a leading producer.

seaweed contains lots of iodine.

BORN FROM SEAWEED AND RAISED IN THE CHIBA PREFECTURE

Iodine is a vital mineral that can be found in our thyroid gland hormones. The Minami Kanto gas field in the Chiba prefecture is one of the largest producers of iodine in the world, second only to Chile. The silver iodide compound can be used in a process called *cloud seeding* to artificially produce rain. This method was actually used in Tokyo during the bone-dry summers of 1996 and 2001.

[áiədàin]
DISCOVERY YEAR: 1811

MELTING POINT
113.6
℃

BOILING POINT
184.4
℃

DENSITY
4.93
g/cm³

| 54 | キセノン
Xenon | 131.3 | 5 / 18 | |

Xe

xenon lamps

Almost as strong as sunlight

Noble gas

Industrial uses

Gaseous

space explorers

Ion engine propellant

THE RISING GAS THAT TRAVELS AMONG THE STARS

[zí:nan]

DISCOVERY YEAR: 1898

The NASA New Millennium program *Deep Space 1* spacecraft, the European Space Agency's *SMART-1*, and the Japanese asteroid probe *Hayabusa* all have one thing in common: Their engines ran on xenon fuel. Xenon engines are about 10 times as fuel effective as their rocket engine counterparts. Xenon is also used as the active gas in plasma displays and as a general anesthetic. Xenon is on the rise!

MELTING POINT
-111.9
℃

BOILING POINT
-107.1
℃

DENSITY
0.0058971
(GAS FORM, 20ºC)
g/cm³

周期

PERIOD

6

原子番号

ATOMIC NUMBER

55→86

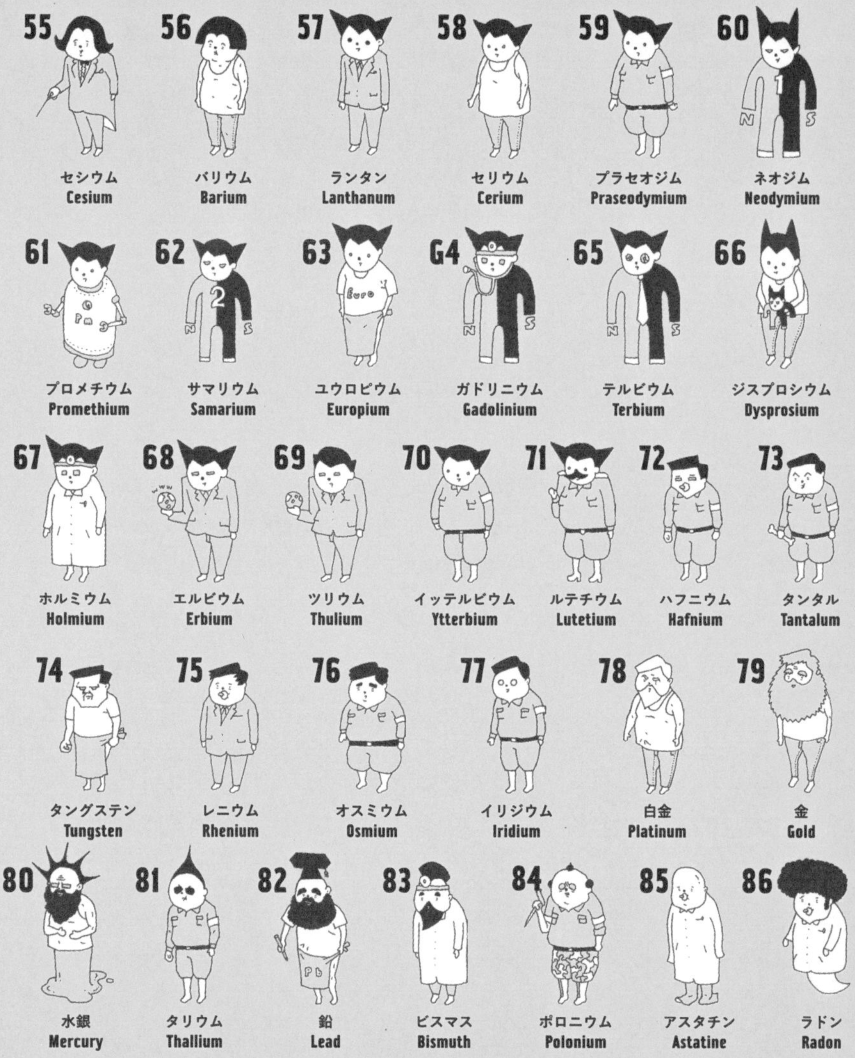

55
セシウム
Cesium
56
バリウム
Barium
57
ランタン
Lanthanum
58
セリウム
Cerium
59
プラセオジム
Praseodymium
60
ネオジム
Neodymium
61
プロメチウム
Promethium
62
サマリウム
Samarium
63
ユウロピウム
Europium
64
ガドリニウム
Gadolinium
65
テルビウム
Terbium
66
ジスプロシウム
Dysprosium
67
ホルミウム
Holmium
68
エルビウム
Erbium
69
ツリウム
Thulium
70
イッテルビウム
Ytterbium
71
ルテチウム
Lutetium
72
ハフニウム
Hafnium
73
タンタル
Tantalum
74
タングステン
Tungsten
75
レニウム
Rhenium
76
オスミウム
Osmium
77
イリジウム
Iridium
78
白金
Platinum
79
金
Gold
80
水銀
Mercury
81
タリウム
Thallium
82
鉛
Lead
83
ビスマス
Bismuth
84
ポロニウム
Polonium
85
アスタチン
Astatine
86
ラドン
Radon

55	セシウム **Cesium**		132.9	6	铯
				—	
				1	

Cs

The period of its electromagnetic wave * 9,192,631,770 = 1 second

Alkali metal

Industrial uses

Solid

Japan's standard time runs on a cesium-based atomic clock.

Elements are rhythmic!

SECOND TO NONE

Have you ever wondered why one second is one second long? Earth's rotational speed was used until 1967, when the General Conference on Weights and Measures decided that the second should be further defined. This is when cesium came into the picture. Now the second is a multiple of the period of cesium's electromagnetic wave. Atomic clocks based on this measurement miss only one second every 1.4 million years.

[síːziəm]
DISCOVERY YEAR: 1860

MELTING POINT
28.40 ℃

BOILING POINT
668.5 ℃

DENSITY
1.873 g/cm³

x

| 56 | バリウム
Barium | 137.3 | 6 / 2 | 钡 |

Alkaline earth metal

Ba

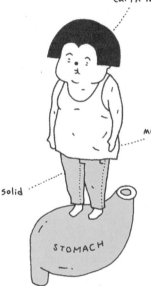

multipurpose

solid

STOMACH

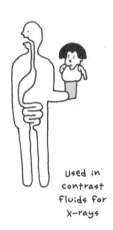

Used in contrast fluids for X-rays

stops X-rays

**A DOCTOR AT WORK,
A GANGSTER AT HOME**

[bέəriəm]
DISCOVERY YEAR: 1808

The white liquid you have to drink before some X-ray procedures is a solution consisting of a powder called barium sulfate and water. It's perfect for analyzing the gastrointestinal tract because X-rays won't pass through it. However, dissolving barium ions in water creates a very strong poison that causes vomiting and paralysis. Pure metallic barium reacts violently when exposed to air, so it's usually preserved in oil.

MELTING POINT
729
℃

BOILING POINT
1637
℃

DENSITY
3.594
g/cm³

| 57 | ランタン
Lanthanum | 138.9

● | 6
—
3 | 鑭 |

La

Lanthanide

Industrial uses

Solid

Used in telescope lenses

LaNi₅

An alloy that absorbs hydrogen

mobile camera lenses

THE LEADER OF THE OUTSIDERS

[lǽnθənəm]

DISCOVERY YEAR: 1839

The next 14 elements are all similar to lanthanum in both their properties and application areas, which is why they (and lanthanum) are grouped together as the lanthanide family. Though some of the other lanthanides are magnetic, lanthanum isn't. It's used as the flint in lighters, in the lenses of mobile cameras, and as a medication to help prevent renal failure.

MELTING POINT
921 ℃

BOILING POINT
3457 ℃

DENSITY
6.145
(25℃)
g/cm³

| 58 | セリウム
Cerium | | 59 | プラセオジム
Praseodymium |

〔síəriəm〕
DISCOVERY YEAR: 1803

Ce

- Lanthanide
- Daily uses
- solid

THE MAINSTAY OF THE LANTHANIDES

铈

140 .1	6	MELTING POINT 799 ℃
	3	BOILING POINT 3426 ℃
		DENSITY (SOLID) (25℃) 6.749 g/cm³

More naturally abundant than copper or silver, cerium is used in sunglasses and UV-resistant glass for its ability to absorb ultraviolet rays. It's also used in engines as a purification catalyst.

〔prèizioudímiəm〕
DISCOVERY YEAR: 1885

Pr

- Lanthanide
- Specialist uses
- solid

THE FLAMING YELLOW MAGICIAN

镨

140 .9	6	MELTING POINT 931 ℃
	3	BOILING POINT 3512 ℃
		DENSITY 6.773 g/cm³

Pure praseodymium is a silver-white solid, but it turns yellow when oxidized. It's often used in welding goggles because it absorbs blue light. Its beautiful yellow is also used in pottery enamel.

60	ネオジム **Neodymium**	144.2	6 / 3	钕

Nd

Lanthanide

solid ..... Magnetic

N S

Hybrid car motors

smack

MRI magnets

mobile phone vibrators

shake   shake

THE WORLD'S STRONGEST SUPER MAGNET

[niːoudímiəm]

DISCOVERY YEAR: 1885

The twin brother of praseodymium was found in the same piece of rock and was consequently named neodymium, which means "the new twin." But one should not take the younger twin lightly! Neodymium, when alloyed with iron and a few other elements, produced the world's strongest magnet in 1982. This new type of magnet was about 1.5 times as strong as the previous record holder and became instantly famous.

MELTING POINT
1021 ℃

BOILING POINT
3068 ℃

DENSITY
7.007 g/cm³

61	プロメチウム **Promethium**

〔prəmíːθiəm〕
DISCOVERY YEAR: 1926

Pm

Lanthanide

man-made

solid

**THE FIERY CHILD
BORN IN OUR REACTORS**

鉅

[145]	6	MELTING POINT 1168 ℃
		BOILING POINT APPROX. 2727 ℃
	3	DENSITY 7.22 g/cm³

The only man-made radioactive lanthanide element is named after the Titan who gave humanity fire: Prometheus. Born in our atomic reactors, it produces heat that's perfect for powering nuclear cells.

62	サマリウム **Samarium**

〔səméəriəm〕
DISCOVERY YEAR: 1879

Sm

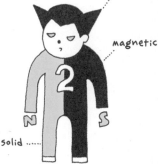

Lanthanide

magnetic

solid

**NUMBER TWO IN THE WORLD
OF MAGNETISM**

釤

150 .4	6	MELTING POINT 1077 ℃
		BOILING POINT 1791 ℃
	3	DENSITY 7.52 g/cm³

The samarium-cobalt magnet was champion before neodymium claimed the title of world's strongest magnet. Even small lanthanide magnets are exceptionally strong, so they're often used in earphones.

| 63 | ユウロピウム
Europium | 152.0 | 6 / 3 | 铕 |

Eu

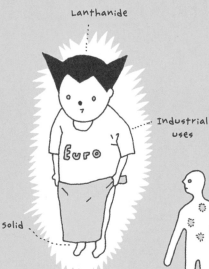

Lanthanide

Industrial uses

Solid

The red display elements of CRT screens

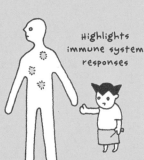

Highlights immune system responses

In luminescent paint

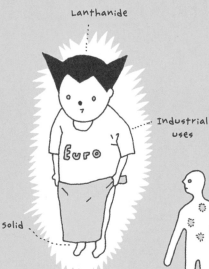

A RESIDENT OF THE NIGHT, LIGHTING UP THE DARK

[juəróupiəm]
DISCOVERY YEAR: 1896

It's the element glowing faintly inside watches and alarm clocks everywhere. It's also used in luminous paint and as an anticounterfeiting measure in euro banknotes. (How appropriate!) But most of the world's europium comes from the US and China. Europium is also in charge of the red component in fluorescent lights and the red display elements in CRT TVs.

MELTING POINT
822
℃

BOILING POINT
1597
℃

DENSITY
5.243
g/cm³

64	ガドリニウム
	Gadolinium

[g`ædəlíniəm]
DISCOVERY YEAR: 1886

Gd

Lanthanide

magnetic

solid

FINDING ILLNESS WITH THE HELP OF MAGNETISM!

釓

157.3	6	**MELTING POINT** 1313 ℃
		BOILING POINT 3266 ℃
	3	**DENSITY (25°C)** 7.9004 g/cm³

Gadolinium is a component of the contrast agent used in most MRI examinations, and it's also in nuclear reactors because of its ability to absorb emitted neutrons well.

65	テルビウム
	Terbium

[té:rbiəm]
DISCOVERY YEAR: 1843

Tb

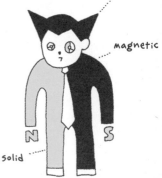

Lanthanide

magnetic

solid

THE OVERLOOKED MAGNET OF YESTERYEAR

158.9	6	**MELTING POINT** 1356 ℃
		BOILING POINT 3123 ℃
	3	**DENSITY** 8.229 g/cm³

Terbium is used in actuators, sonar systems, and fluorescent lamps. It's also used in electric bicycles and magnetic glass due to its magnetic properties.

66 ジスプロシウム
Dysprosium

[dispróusiəm]
DISCOVERY YEAR: 1886

Dy

Lanthanide

Daily uses

solid

**THE STRONGEST TAG TEAM!
DYSPROSIUM AND NEODYMIUM**

镝

162 .5	6	MELTING POINT 1412 ℃
		BOILING POINT 2562 ℃
	3	DENSITY 8.55 g/cm³

Even the strongest neodymium magnet weakens when heated. That's where dysprosium comes in. This combination is essential in places where high temperatures are the norm, like hybrid car engines.

---

67 ホルミウム
Holmium

[hóulmiəm]
DISCOVERY YEAR: 1879

Ho

Lanthanide

scientific uses

solid

**A PAL TO PROSTATES
EVERYWHERE**

钬

164 .9	6	MELTING POINT 1474 ℃
		BOILING POINT 2395 ℃
	3	DENSITY 8.795 g/cm³

Holmium lasers are a perfect treatment method for prostatic hypertrophy. The laser prevents hemorrhage as the incision is performed. It is also great for removing renal and urethral stones.

68	エルビウム **Erbium**

【éːrbiəm】
DISCOVERY YEAR: 1843

Er

Lanthanide

Industrial uses

www

solid

MANAGING OUR WORLDWIDE NETWORKS

铒

167.3	6	**MELTING POINT** 1529 ℃
		BOILING POINT 2863 ℃
	3	**DENSITY (25℃)** **9.066** g/cm³

When we send data over the Internet, we're sending it as light pulses through long, reflecting cables; doing this over long distances would be impossible without erbium light-amplification relays.

69	ツリウム **Thulium**

【θjúːliəm】
DISCOVERY YEAR: 1879

Tm

Lanthanide

Industrial uses

solid

ERBIUM'S LITTLE BROTHER

铥

168.9	6	**MELTING POINT** 1545 ℃
		BOILING POINT 1947 ℃
	3	**DENSITY** **9.321** g/cm³

Thulium is still not used much in industry due to being very rare and very hard to isolate. It is, however, much like erbium, used in optic fiber light-amplification units.

70 イッテルビウム
Ytterbium

[itéːrbiəm]
DISCOVERY YEAR: 1878

Yb

Lanthanide

Specialist
uses

Solid

**ANOTHER ONE FROM
TEAM SCANDINAVIA**

鐿

173 .0	6	MELTING POINT 824 ℃
		BOILING POINT 1193 ℃
	3	DENSITY 6.965 g/cm³

Its name comes from Ytterby, a small town in Sweden where a multitude of elements have been discovered. Ytterbium's uses are very similar to those of erbium, and it can color glass yellow–green.

71 ルテチウム
Lutetium

[luːtíːʃiəm]
DISCOVERY YEAR: 1907

Lu

Lanthanide

Specialist
uses

Solid

**MORE EXPENSIVE THAN GOLD!
THE ROYAL ELEMENT**

镥

175 .0	6	MELTING POINT 1663 ℃
		BOILING POINT 3395 ℃
	3	DENSITY 9.84 g/cm³

It's hard to believe, but lutetium costs a whopping ¥50,500* per gram! That's more than the price of silver, gold, and platinum combined. It doesn't really have any applications outside of research, though.

* There are roughly 100 Japanese yen to 1 US dollar.

72 ハフニウム
Hafnium

[hǽfniəm]
DISCOVERY YEAR: 1922

Hf

Transition metal

Specialist uses

solid

ZIRCONIUM'S SIGNIFICANT OTHER

铪

178 .5	6	MELTING POINT 2230 ℃
		BOILING POINT 5197 ℃
	4	DENSITY 13.31 g/cm³

With properties very similar to zirconium's, hafnium is sometimes used in nuclear reactor control rods to absorb neutrons, while zirconium takes the opposite role of the reactor's fuel rods.

73 タンタル
Tantalum

[tǽntələm]
DISCOVERY YEAR: 1802

Ta

Transition metal

Specialist uses

solid

FOR BONE PROSTHESES AND MOBILE PHONES

钽

180 .9	6	MELTING POINT 2996 ℃
		BOILING POINT 5425 ℃
	5	DENSITY 16.654 g/cm³

Since the human body tolerates tantalum well, it is often used for bone prostheses, artificial joints, and dental implants. It's also used in small, efficient electric capacitors for mobile phones and laptops.

| 74 | タングステン
Tungsten | 183.8
・ | 6
—
6 | 钨 |

W

Transition metal

Daily uses

Solid

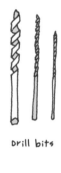

Drill bits

As the filament in lightbulbs

Forms extremely strong steel with carbon

THE WORLD'S MOST THICK-SKINNED ARTISAN

When Edison invented the light bulb, he used a piece of wick as his filament, but it burned too fast to be useful and broke easily. In the 20th century, we began using tungsten to make filaments, and thus the tungsten halogen lamp was born. Tungsten has the highest melting point of all the elements. When carbonized, it produces a super material that's almost as hard as diamond and is used to make abrasion-resistant drills and molds.

[tʌ́ŋstən]

DISCOVERY YEAR: 1781

MELTING POINT
3407
℃

BOILING POINT
5657
℃

DENSITY
19.3
g/cm³

75	レニウム
	Rhenium

[ríːniəm]
DISCOVERY YEAR: 1925

Re

Transition metal

Industrial uses

Solid

**OUR MOST RECENT
NATURAL FIND**

186 .2	6	**MELTING POINT** 3180 ℃
		BOILING POINT 5627 ℃
	7	**DENSITY** 21.02 g/cm³

Rhenium is our most recent natural find. It has the second-highest melting point, just below that of tungsten. This makes it ideal for high-temperature measuring equipment and rocket nozzles.

76	オスミウム
	Osmium

[ázmiəm]
DISCOVERY YEAR: 1803

Os

Transition metal

Specialist uses

Solid

**THE HEAVIEST SUMO
OF THEM ALL**

190 .2	6	**MELTING POINT** 3054 ℃
		BOILING POINT 5027 ℃
	8	**DENSITY** 22.59 g/cm³

The densest element and the heaviest metal, osmium becomes very abrasion- and rust-resistant when alloyed with iridium, ruthenium, and platinum. Its durability suits it for fountain pen tips.

| 77 | イリジウム
Iridium | 192.2 | 6 / 9 | 銥 |

Ir

Transition metal

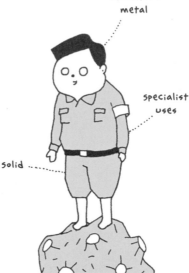

Specialist uses

solid

Iridium deposits in Earth's crust support the theory that the extinction of the dinosaurs was caused by a meteorite.

1m

Until 1960, the international prototype meter was made out of a platinum and iridium alloy.

spark plugs are made of iridium alloys.

THE ELEMENT CLOSEST TO ETERNITY

[irídiəm]
DISCOVERY YEAR: 1803

Gold and platinum are well known for being used to make wedding rings and other jewelry because of their nonreactive natures, but the most resilient metal of all is actually iridium. Because of this, the international prototype kilogram is made of an alloy of about 10% iridium and 90% platinum, as was the international prototype meter until 1960. If you would like to swear an oath for eternal love, iridium might be your best bet.

MELTING POINT	**2410** ℃
BOILING POINT	**4130** ℃
DENSITY	**22.56** g/cm³

| 78 | 白金 (プラチナ)
 Platinum | 195.1
 • | 6
 —
 10 | 铂 |

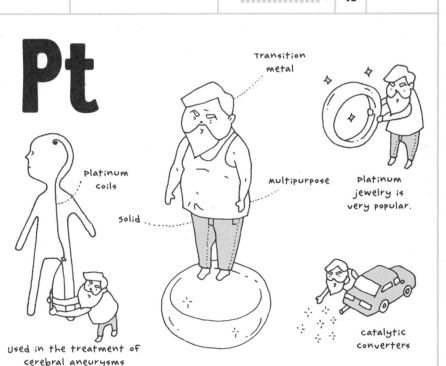

Pt

Transition metal

platinum coils

solid

multipurpose

Platinum jewelry is very popular.

Used in the treatment of cerebral aneurysms

catalytic converters

THE LATE-BLOOMING STAR

[plǽtənəm]
DISCOVERY YEAR: 1751

Platinum is popular now, but it played second fiddle to its older siblings gold and silver when it was discovered in the 18th century. Its name even means "small silver" in Spanish (*platina*). But today, due to its exceptional corrosion resistance, it's used in jewelry, electrodes in physical and chemical science, and coils for treating cerebral aneurisms. It's also a key part of some cancer-fighting drugs.

MELTING POINT
1772
℃

BOILING POINT
3827
℃

DENSITY
21.45
g/cm³

| 79 | 金
Gold | 197.0 | 6 | 金 |
| | | | 11 | |

False teeth

Au

Transition
metal

very
malleable

money

multipurpose

solid

Heh
heh

I'm in
love!

GOLD
GOLD 24

THE SYMBOL OF PROSPERITY, WEALTH, AND POWER

[góuld]
DISCOVERY YEAR: ANCIENT

Gold has always been a symbol of power, from King Tutankhamun's golden mask to the gleaming teeth of hip-hop mainstay Flavor Flav. In the Middle Ages, alchemists tried to create gold from other metals; their efforts served as a precursor to modern chemistry. Gold is also used in circuitry because of its excellent heat and electrical conductivity and in medals and coins for its beauty and corrosion resistance.

MELTING POINT
1064.43
℃

BOILING POINT
2807
℃

DENSITY
19.32
g/cm³

| 80 | 水銀
Mercury | 200.6 | 6 | 汞 |
| | | | 12 | |

Hg

The zinc family

Roll roll

very high surface tension

multipurpose

Liquid

In older thermometers

Poisonous

THE MUTANT OF THE METAL WORLD

[méːrkjuri]

DISCOVERY YEAR: ANCIENT

Mercury is the only metal to be in liquid form and capable of evaporating at room temperature. It creates soft alloys (amalgams) when combined with other metals and has been used as plating for many years. It is still popular in thermometers and mercury vapor lamps. It is important to remember that while it may be easy to work with, it is highly toxic and can become a double-edged sword if one is not careful.

MELTING POINT
-38.87
℃

BOILING POINT
356.58
℃

DENSITY
13.546
(LIQUID, 20ºC)
g/cm³

81	タリウム **Thallium**	204.4 6 / 13	鉈

TI

The boron family

Specialist uses

Like butter

throb throb

solid

Used in nuclear cardiography

soft metal

The most toxic of all the heavy metals

WITH THE UNEXPECTED ABILITY TO DETECT HEART ATTACKS

[θǽliəm]
DISCOVERY YEAR: 1861

Thallium is known for being almost as toxic as arsenic. A single gram is enough to kill an adult. It was the British serial killer Graham Young's murder weapon of choice and also appeared in Agatha Christie's *The Pale Horse*. It was also widely used as a rat and ant poison until the 1970s, when this use was prohibited for obvious reasons. More helpfully, it is used as a radioactive isotope to help us find irregular blood flows and the like.

MELTING POINT
303.5 ℃

BOILING POINT
1457 ℃

DENSITY
11.85 g/cm³

82

鉛
Lead

207.2

6
—
14

铅

Pb

Doesn't let
radiation through

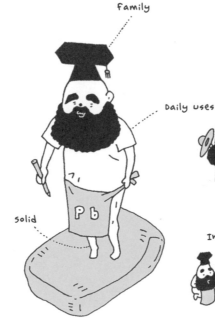

The carbon
family

Daily uses

Solid

Pb

Fishing
sinkers

In solder

CAR
BATTERY

HANDA

**THE WORLD AUTHORITY
WHO WAS FORCED INTO
EARLY RETIREMENT**

[led]
DISCOVERY YEAR: ANCIENT

Lead is easy to work with and has had many uses over the years. The ancient Romans used it to build their waterways, but since it's a strong poison, that might have played a role in the fall of the Roman Empire. The word *plumbing* and the abbreviation *Pb* come from the Latin word for lead. Modern uses include car batteries, solder, and mirrors, but because of its toxicity and limited reserves, lead is being phased out of many applications.

MELTING POINT
327.50
℃

BOILING POINT
1740
℃

DENSITY
11.35
g/cm³

151

83 ビスマス
Bismuth

Bi

[bízməθ]
DISCOVERY YEAR: 1753

The nitrogen family

Daily uses

solid

LEAD'S FAITHFUL SUCCESSOR

秘

209 .0	6	MELTING POINT 271.3 ℃
	---	---
	—	BOILING POINT 1560 ℃
	15	DENSITY 9.747 g/cm³

Bismuth is useful both in alloys and in medical applications, such as remedies for gastric ulcers and diarrhea. Since it's similar to lead, it's gaining popularity as a nontoxic lead replacement.

84 ポロニウム
Polonium

Po

[pəlóuniəm]
DISCOVERY YEAR: 1898

Radioactive

The oxygen family

Specialist uses

solid

THE MOST DESTRUCTIVE OF THE NATURAL ELEMENTS

釙

[210]	6	MELTING POINT 254 ℃
	---	---
	—	BOILING POINT 962 ℃
	16	DENSITY 9.32 g/cm³

The naturally radioactive element polonium was the first element to be discovered by the Curies, with a radioactive intensity about 330 times as strong as that of uranium.

85 アスタチン
Astatine

【ǽstətíːn】
DISCOVERY YEAR: 1940

At

- Radioactive
- Halogen
- Scientific uses
- Solid

THE LAST SAMURAI OF THE HALOGENS

砹

[210]	6	MELTING POINT	302 ℃
	---	BOILING POINT	337 ℃
	17	DENSITY	... g/cm³

Naturally occurring astatine is the most rarely encountered element in nature and has to be synthesized in order to be studied. Determining its properties is very hard because its half-life is so short.

86 ラドン
Radon

【réidan】
DISCOVERY YEAR: 1900

Rn

- Noble gas
- Radioactive
- Scientific uses
- Gaseous

THE CHUBBY BATHING BEAUTY

[222]	6	MELTING POINT	-71 ℃
	---	BOILING POINT	-61.8 ℃
	18	DENSITY (GAS, 0℃)	0.00973 g/cm³

Radon is the heaviest gaseous element at room temperature. Hot springs containing radon are said to have a positive effect on any bather's health, but breathing radon can cause lung cancer.

153

周期
PERIOD

7

原子番号

ATOMIC NUMBER

87→118

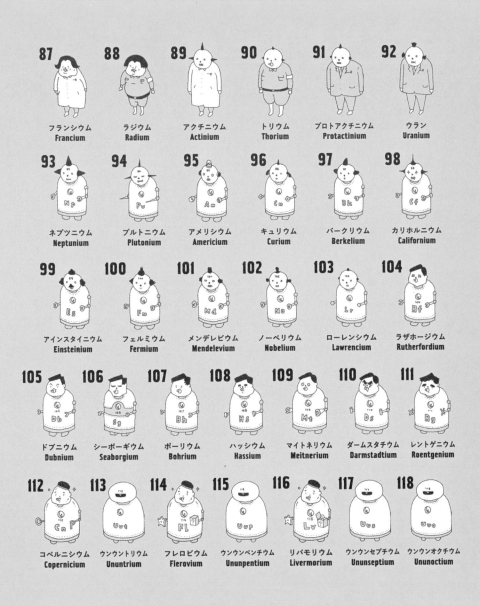

87	88	89	90	91	92
フランシウム	ラジウム	アクチニウム	トリウム	プロトアクチニウム	ウラン
Francium	**Radium**	**Actinium**	**Thorium**	**Protactinium**	**Uranium**

93	94	95	96	97	98
ネプツニウム	プルトニウム	アメリシウム	キュリウム	バークリウム	カリホルニウム
Neptunium	**Plutonium**	**Americium**	**Curium**	**Berkelium**	**Californium**

99	100	101	102	103	104
アインスタイニウム	フェルミウム	メンデレビウム	ノーベリウム	ローレンシウム	ラザホージウム
Einsteinium	**Fermium**	**Mendelevium**	**Nobelium**	**Lawrencium**	**Rutherfordium**

105	106	107	108	109	110	111
ドブニウム	シーボーギウム	ボーリウム	ハッシウム	マイトネリウム	ダームスタチウム	レントゲニウム
Dubnium	**Seaborgium**	**Bohrium**	**Hassium**	**Meitnerium**	**Darmstadtium**	**Roentgenium**

112	113	114	115	116	117	118
コペルニシウム	ウンウントリウム	フレロビウム	ウンウンペンチウム	リバモリウム	ウンウンセプチウム	ウンウンオクチウム
Copernicium	**Ununtrium**	**Flerovium**	**Ununpentium**	**Livermorium**	**Ununseptium**	**Ununoctium**

87	フランシウム
	Francium

[frǽnsiəm]
DISCOVERY YEAR: 1939

Fr

- Alkali metal
- Radioactive
- Scientific uses
- solid

THE FLEETING MYSTERY

钫

[223]	7	MELTING POINT	27 ℃
		BOILING POINT	677 ℃
	1	DENSITY	--- g/cm³

Francium has the shortest half-life of all naturally occurring radioactive elements at about 22 minutes. It is thought that the element is solid at room temperature, but that is still under debate.

88	ラジウム
	Radium

[réidiəm]
DISCOVERY YEAR: 1898

Ra

- Alkaline earth metal
- Radioactive
- specialist uses
- solid

THE ELEMENT THAT BIT THE HAND THAT FED IT

镭

[226]	7	MELTING POINT	700 ℃
		BOILING POINT	1140 ℃
	2	DENSITY	APPROX. 5 g/cm³

This element was discovered by Marie Curie in 1898. She received the Nobel prize in chemistry 1911 for her work but died a few decades later from ailments brought on by prolonged exposure to radiation.

89	アクチニウム
	Actinium

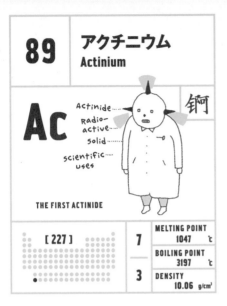

Ac

Actinide
Radio-active
Solid
Scientific uses

銅

THE FIRST ACTINIDE

	7	MELTING POINT 1047 ℃
[227]		BOILING POINT 3197 ℃
	3	DENSITY 10.06 g/cm³

91	プロトアクチニウム
	Protactinium

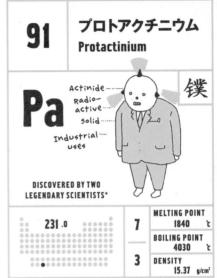

Pa

Actinide
Radio-active
Solid
Industrial uses

鏷

DISCOVERED BY TWO LEGENDARY SCIENTISTS*

	7	MELTING POINT 1840 ℃
231 .0		BOILING POINT 4030 ℃
	3	DENSITY 15.37 g/cm³

* Germany's Otto Haan and Lise Meitner

90	トリウム
	Thorium

Th

Actinide
Radio-active
Solid
Specialist uses

釷

HOLDS GREAT PROMISE AS THE FUEL OF TOMORROW

	7	MELTING POINT 1750 ℃
232 .0		BOILING POINT 4787 ℃
	3	DENSITY 11.72 g/cm³

92	ウラン
	Uranium

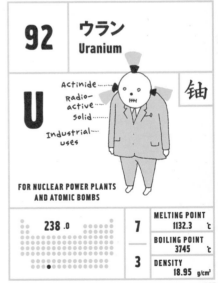

U

Actinide
Radio-active
Solid
Industrial uses

铀

FOR NUCLEAR POWER PLANTS AND ATOMIC BOMBS

	7	MELTING POINT 1132.3 ℃
238 .0		BOILING POINT 3745 ℃
	3	DENSITY 18.95 g/cm³

93 ネプツニウム
Neptunium

Np 鎿

Actinide
Radioactive
Solid
man-made

EVEN HEAVIER THAN URANIUM

[237]

7
—
3

MELTING POINT
640 ℃

BOILING POINT
3902 ℃

DENSITY
20.25 g/cm³

95 アメリシウム
Americium

Am 镅

Actinide
Radio-active
Solid
man-made

USED IN SMOKE DETECTORS

[243]

7
—
3

MELTING POINT
1172 ℃

BOILING POINT
2607 ℃

DENSITY
13.67 g/cm³

94 プルトニウム
Plutonium

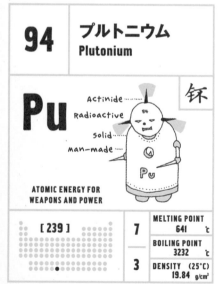

Pu 钚

Actinide
Radioactive
Solid
man-made

ATOMIC ENERGY FOR WEAPONS AND POWER

[239]

7
—
3

MELTING POINT
641 ℃

BOILING POINT
3232 ℃

DENSITY (25°C)
19.84 g/cm³

96 キュリウム
Curium

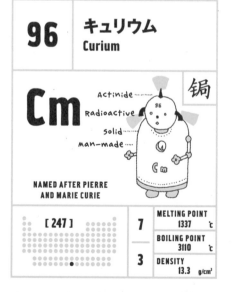

Cm 锔

Actinide
Radioactive
Solid
man-made

NAMED AFTER PIERRE AND MARIE CURIE

[247]

7
—
3

MELTING POINT
1337 ℃

BOILING POINT
3110 ℃

DENSITY
13.3 g/cm³

97 バークリウム / Berkelium

Bk

铝

Actinide
Radioactive
Solid
man-made

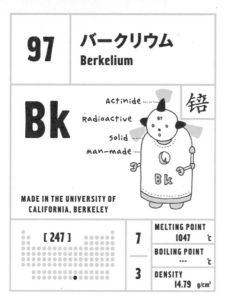

MADE IN THE UNIVERSITY OF
CALIFORNIA, BERKELEY

[247]

7

3

MELTING POINT	1047 ℃
BOILING POINT	--- ℃
DENSITY	14.79 g/cm³

99 アインスタイニウム / Einsteinium

Es

锿

Actinide
Radioactive
Solid
man-made

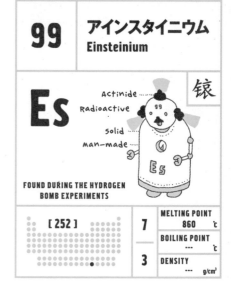

FOUND DURING THE HYDROGEN
BOMB EXPERIMENTS

[252]

7

3

MELTING POINT	860 ℃
BOILING POINT	--- ℃
DENSITY	--- g/cm³

98 カリホルニウム / Californium

Cf

锎

Actinide
Radioactive
Solid
man-made

IT'S SUPER EXPENSIVE!
ONE GRAM COSTS
A BILLION DOLLARS?!

[252]

7

3

MELTING POINT	897 ℃
BOILING POINT	--- ℃
DENSITY	15.1 g/cm³

100 フェルミウム / Fermium

Fm

镄

Actinide
Radioactive
Solid
man-made

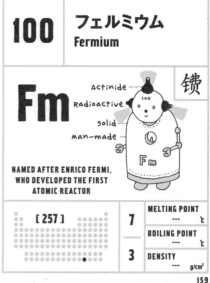

NAMED AFTER ENRICO FERMI,
WHO DEVELOPED THE FIRST
ATOMIC REACTOR

[257]

7

3

MELTING POINT	--- ℃
BOILING POINT	--- ℃
DENSITY	--- g/cm³

101 メンデレビウム
Mendelevium

Md
钔

Actinide
Radioactive
Solid
man-made

NAMED AFTER THE FATHER
OF THE TABLE OF THE
ELEMENTS, MENDELEEV

[258]

7
3

MELTING POINT	--- ℃
BOILING POINT	--- ℃
DENSITY	--- g/cm³

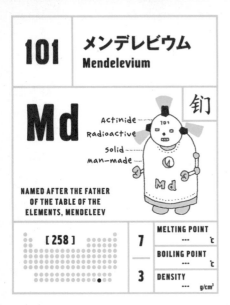

103 ローレンシウム
Lawrencium

Lr
铹

Actinide
Radioactive
Solid
man-made

NAMED AFTER ERNEST
LAWRENCE, THE PHYSICIST

[262]

7
3

MELTING POINT	--- ℃
BOILING POINT	--- ℃
DENSITY	--- g/cm³

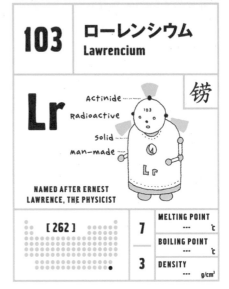

102 ノーベリウム
Nobelium

No
锘

Actinide
Radioactive
Solid
man-made

NAMED AFTER THE
HONORABLE ALFRED NOBEL

[259]

7
3

MELTING POINT	--- ℃
BOILING POINT	--- ℃
DENSITY	--- g/cm³

104 ラザホージウム
Rutherfordium

Rf
𬬻

Transition
metal
Radioactive
Solid
man-made

NAMED AFTER ERNEST
RUTHERFORD, WHO DISCOVERED
THE STRUCTURE OF THE ATOM

[267]

7
4

MELTING POINT	--- ℃
BOILING POINT	--- ℃
DENSITY	23 g/cm³

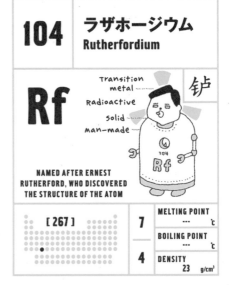

105 ドブニウム
Dubnium

Db

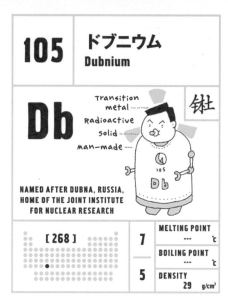

鉨

Transition metal
Radioactive
Solid
man-made

NAMED AFTER DUBNA, RUSSIA,
HOME OF THE JOINT INSTITUTE
FOR NUCLEAR RESEARCH

[268]

7

5

MELTING POINT
--- ℃

BOILING POINT
--- ℃

DENSITY
29 g/cm³

107 ボーリウム
Bohrium

Bh

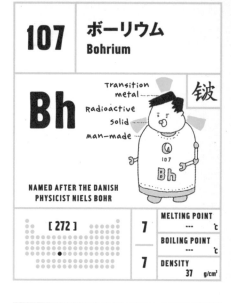

𨨏

Transition metal
Radioactive
Solid
man-made

NAMED AFTER THE DANISH
PHYSICIST NIELS BOHR

[272]

7

7

MELTING POINT
--- ℃

BOILING POINT
--- ℃

DENSITY
37 g/cm³

106 シーボーギウム
Seaborgium

Sg

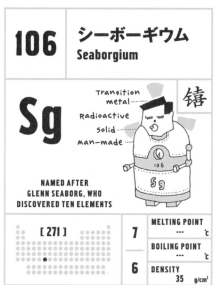

𨭎

Transition metal
Radioactive
Solid
man-made

NAMED AFTER
GLENN SEABORG, WHO
DISCOVERED TEN ELEMENTS

[271]

7

6

MELTING POINT
--- ℃

BOILING POINT
--- ℃

DENSITY
35 g/cm³

108 ハッシウム
Hassium

Hs

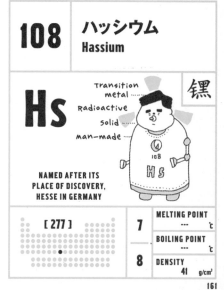

𨭆

Transition metal
Radioactive
Solid
man-made

NAMED AFTER ITS
PLACE OF DISCOVERY,
HESSE IN GERMANY

[277]

7

8

MELTING POINT
--- ℃

BOILING POINT
--- ℃

DENSITY
41 g/cm³

109 マイトネリウム
Meitnerium

Mt

Transition metal
Radioactive
Solid
man-made

锿

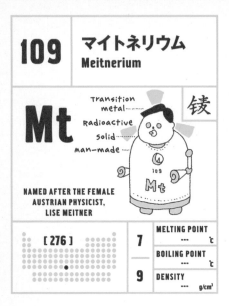

NAMED AFTER THE FEMALE
AUSTRIAN PHYSICIST,
LISE MEITNER

[276]

7	MELTING POINT --- ℃
	BOILING POINT --- ℃
9	DENSITY --- g/cm³

111 レントゲニウム
Roentgenium

Rg

Transition metal
Radioactive
Solid
man-made

铑

NAMED AFTER THE PHYSICIST
WHO DISCOVERED THE X-RAY,
WILHELM RÖNTGEN

[280]

7	MELTING POINT --- ℃
	BOILING POINT --- ℃
11	DENSITY --- g/cm³

110 ダームスタチウム
Darmstadtium

Ds

Transition metal
Radioactive
Solid
man-made

𫟼

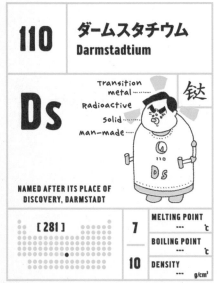

NAMED AFTER ITS PLACE OF
DISCOVERY, DARMSTADT

[281]

7	MELTING POINT --- ℃
	BOILING POINT --- ℃
10	DENSITY --- g/cm³

112 コペルニシウム
Copernicium

Cn

Radioactive

man-made

鿔

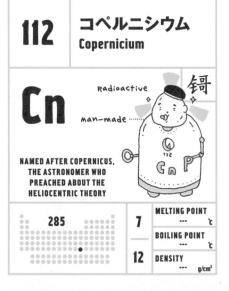

NAMED AFTER COPERNICUS,
THE ASTRONOMER WHO
PREACHED ABOUT THE
HELIOCENTRIC THEORY

285

7	MELTING POINT --- ℃
	BOILING POINT --- ℃
12	DENSITY --- g/cm³

113 ウンウントリウム Ununtrium

Uut

284

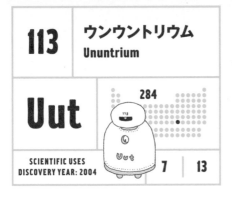

SCIENTIFIC USES
DISCOVERY YEAR: 2004

7 | 13

116 リバモリウム Livermorium

Lv

Radioactive

293

man-
made

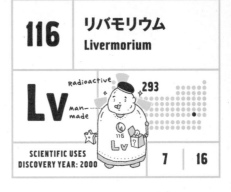

SCIENTIFIC USES
DISCOVERY YEAR: 2000

7 | 16

114 フレロビウム Flerovium

Fl

Radioactive

289

man-
made

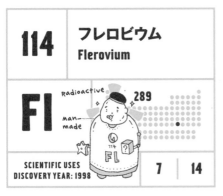

SCIENTIFIC USES
DISCOVERY YEAR: 1998

7 | 14

117 ウンウンセプチウム Ununseptium

Uus

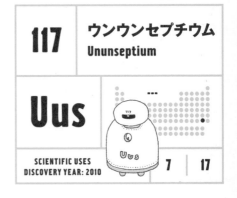

SCIENTIFIC USES
DISCOVERY YEAR: 2010

7 | 17

115 ウンウンペンチウム Ununpentium

Uup

288

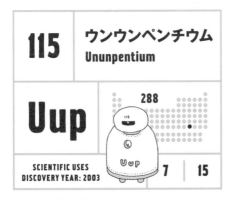

SCIENTIFIC USES
DISCOVERY YEAR: 2003

7 | 15

118 ウンウンオクチウム Ununoctium

Uuo

294

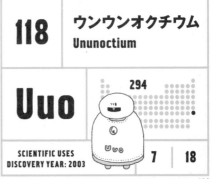

SCIENTIFIC USES
DISCOVERY YEAR: 2003

7 | 18

ELEMENT PRICE RANKINGS*

These are the top five elements that are sold as reagents. Elements come in all different shapes and colors, so it's hard to make any generalizations. This list is based on 1 gram samples of all the elements to give you a general feeling of their relative prices. Special elements like uranium and plutonium can't really be evaluated, so they are not listed. Gold and platinum look pretty cheap when put into perspective like this!

1

Rh

RHODIUM

¥60,000

IG POWDER
(99.9% PURE)

2

Cs

CESIUM

¥52,400

IG ENCLOSED SAMPLE

3

Lu

LUTETIUM

¥50,500

IG FRAGMENT
(99.9% PURE)

4

Sc

SCANDIUM

¥45,900

IG INGOT
(99.9% PURE)

5

Tm

THULIUM

¥33,100

IG PELLET

SOME PRECIOUS METALS
FOR COMPARISON

PLATINUM	**¥4,216**
GOLD	**¥3,139**
SILVER	**¥51.6**

* There are roughly 100 Japanese yen to 1 US dollar.

THE COST OF ONE HUMAN BEING

How much does a human cost? I tried to calculate the price using common materials that anyone can buy and included most of the elements in the human body. If we assume that the person weighs around 60 kg (132 lbs), the body's worth roughly ¥13,000. I guess it's up to each person to decide how much tax goes on top of that...

Element	Cost	Equivalent
ZINC	¥0.5	0.12 G EQUIVALENT OF ZINC FOR EXPERIMENTAL USE
IRON	¥14	3 G EQUIVALENT IN IRON NAILS
SODIUM & CHLORINE	¥20	180 G EQUIVALENT IN TABLE SALT
SULFUR	¥288	120 G EQUIVALENT IN SULFUR FOR EXPERIMENTAL USE
PHOSPHORUS	¥300	600 G EQUIVALENT IN PHOSPHORUS-BASED FERTILIZER
POTASSIUM	¥605	240 G EQUIVALENT IN POTASSIUM-BASED FERTILIZER
NITROGEN	¥774	1.8 KG EQUIVALENT IN NITROGEN-BASED FERTILIZER
CARBON	¥896	10.8 KG EQUIVALENT IN BARBECUE COAL
CALCIUM	¥1,766	0.9 KG EQUIVALENT IN CALCIUM CARBONATE FOR EXPERIMENTAL USE
OXYGEN & HYDROGEN	¥3,980	45 KG EQUIVALENT IN WATER
MAGNESIUM	¥4,200	30 G EQUIVALENT IN MAGNESIUM FOR EXPERIMENTAL USE
OTHERS		

≈ ¥13,000

ELEMENT FRIENDS

Among the 118 elements, certain groups of elements have similar properties, and some of them even reinforce each other's reactions. There are elements who play well with others and others who just want to pick a fight...

Na K Rb Cs

THE FOUR EXPLOSIVE ALKALI EMPERORS

These four elements may seem like a peaceful bunch, but if you get them wet, you'll see just how explosive their tempers can be! Their pure forms must be kept submerged in oil to prevent the violent reaction caused by contact with water. From least explosive to most explosive they are Sodium, Potassium, Rubidium, and Cesium.

Au Ag Cu

THE THREE SAGES OF WEALTH AND PROSPERITY

Gold, silver, and copper are all abundant, easy to work with, and corrosion resistant, which makes them an exceptionally accomplished team of metals. This is why they have been used since ancient times as currency, raw materials, and prized possessions. The well-known set of Olympic medals is just one example of many.

Si **Ge** **Sn**

THE DIGITAL SEMICONDUCTOR TRIO

Silicon, germanium, and tin are the three main elements used in semiconductor construction. They are the elite few that helped Japan become one of the leading countries in electronics. It is thanks to them that we have access to computers and other digital devices today.

Nd **Sm**

THE STRONGEST MAGNET COMBO IN THE WORLD

Neodymium and samarium are engaged in an eternal struggle for the title of "world's best magnet." That honor currently goes to neodymium, but samarium magnets are both more heat resistant and more rugged, which makes them the better choice in many applications.

Ca **Sr** **Ba**

THE CASBAH BROTHERS

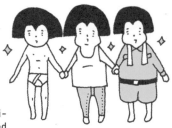

Sometimes elements with very similar properties and very regularly spaced atomic weights form groups of three in the table of elements. These groups are called "triads." Calcium, strontium, and barium form one of these groups, and since their starting letters are *Ca*, *S*, and *Ba*, I thought "the Casbah brothers" might be a good family name for them.

TROUBLESOME ELEMENTS

Elements that aren't that dangerous by themselves can gain unimaginable destructive power when paired with a few others. I thought we could have a look at a few of the groups that have been stirring up trouble in the world these last few decades.

C₂H₈NO₂PS

METHAMIDOPHOS

Methamidophos became famous in Japan when trace amounts of the poison were found in foodstuffs imported from China. It is made up of a multitude of elements.

As₂O₃ (As₄O₆)

ARSENIC TRIOXIDE

Arsenic trioxide was used in the assassination of Napoleon and in the infamous Wakayama curry poisoning in the summer of 1998.

C₄H₁₀O₂FP

SARIN

Even though sarin is made up of some very familiar elements, it is an extremely potent nerve gas.

CH₂O

FORMALDEHYDE

This harmful indoor air pollutant was named as one of the elements responsible for "sick building syndrome" in the 1980s.

KCN

POTASSIUM CYANIDE

The classic poison used throughout history has a surprisingly simple chemical formula.

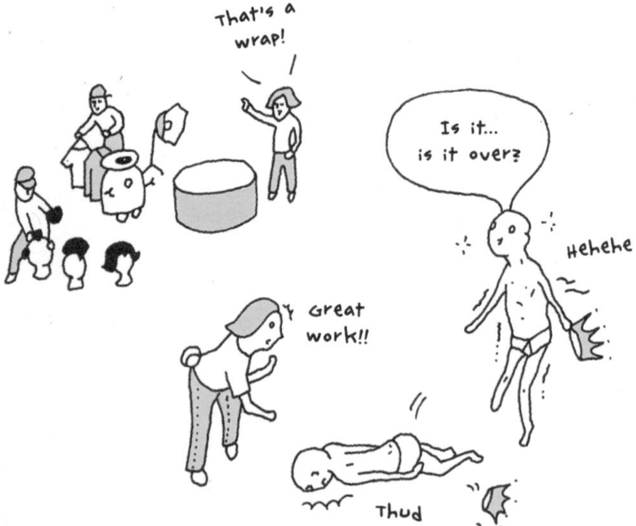
That's a wrap!
Is it... is it over?
Hehehe
Great work!!
Thud

4

HOW TO EAT THE ELEMENTS
元素の食べ方

Our bodies are also made of elements—about 34 different elements, actually. That means that over one third of all the elements we've looked at so far are actually a part of us. It's easy to think that elements exist only in the outside world, but...

WE'RE ALL ELEMENT TREASURE HOUSES.

And among them are lots of elements that you might have thought you'd never have anything to do with, like strontium or molybdenum. It might surprise you to know that arsenic is one of them, too. Arsenic, which is almost synonymous with poison, actually exists naturally within us. This is also true for other unfamiliar elements like cadmium, beryllium, and radium. They're all a part of our bodies.

But of course elements are not created inside our bodies. They are all there because we've eaten them at some point. Before that, they were part of some other entity.

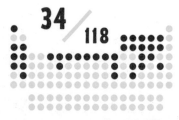

● = ELEMENTS IN OUR BODIES

THE ELEMENTS INSIDE OUR BODIES

 H
HYDROGEN

 B
BORON

 C
CARBON

 N
NITROGEN

 O
OXYGEN

 F
FLUORINE

 Na
SODIUM

 Mg
MAGNESIUM

 Al
ALUMINUM

 Si
SILICON

 P
PHOSPHORUS

 S
SULFUR

 Cl
CHLORINE

 K
POTASSIUM

 Ca
CALCIUM

 V
VANADIUM

 Cr
CHROMIUM

 Mn
MANGANESE

 Fe
IRON

 Co
COBALT

 Ni
NICKEL

 Cu
COPPER

 Zn
ZINC

 As
ARSENIC

 Se
SELENIUM

 Rb
RUBIDIUM

 Sr
STRONTIUM

 Mo
MOLYBDENUM

 Cd
CADMIUM

 Sn
TIN

 I
IODINE

 Ba
BARIUM

Hg
MERCURY

Pb
LEAD

173

The average human is made up of about 65% oxygen, 18% carbon, and 10% hydrogen.

WAIT A SECOND! THAT'S ALMOST 100%!

In reality, about 28 of those 34 elements don't even amount to 1% of our total mass. But just because these elements appear in tiny amounts doesn't mean they're not important—quite the opposite! Even if only a tenth of a percent of the elements in our bodies were to go missing, we'd be dead. These low-volume but important elements are called *trace elements*, and most of them are metals. The most important of these are called...

MINERALS.

Minerals are absolutely necessary to all living beings, including humans.

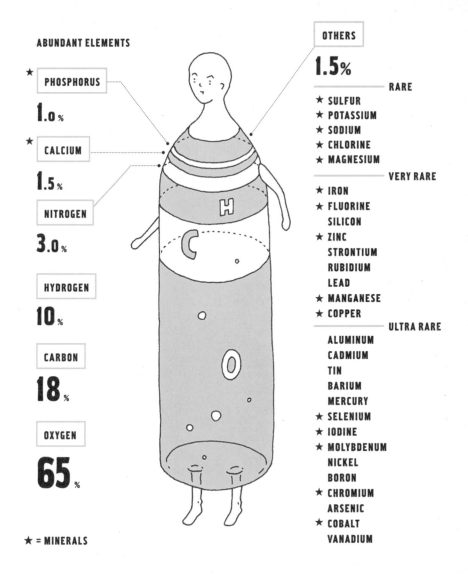

ABUNDANT ELEMENTS

★ PHOSPHORUS
1.0 %

★ CALCIUM
1.5 %

NITROGEN
3.0 %

HYDROGEN
10 %

CARBON
18 %

OXYGEN
65 %

★ = MINERALS

OTHERS
1.5%

————————— RARE
★ SULFUR
★ POTASSIUM
★ SODIUM
★ CHLORINE
★ MAGNESIUM

————————— VERY RARE
★ IRON
★ FLUORINE
  SILICON
★ ZINC
  STRONTIUM
  RUBIDIUM
  LEAD
★ MANGANESE
★ COPPER

————————— ULTRA RARE
  ALUMINUM
  CADMIUM
  TIN
  BARIUM
  MERCURY
★ SELENIUM
★ IODINE
★ MOLYBDENUM
  NICKEL
  BORON
★ CHROMIUM
  ARSENIC
★ COBALT
  VANADIUM

Right now, there are around 17 recognized dietary minerals.* They are the starting point for many compounds, and they help control how other elements react with each other.

THEY ARE LIKE THE PLAYMAKERS OF OUR BODIES.

If the body were an orchestra, the minerals would be its conductor. If it were an airport, the minerals would be its control tower. If a company, its director. That is what minerals do. If we run low on iron, we get anemic, and if we don't get enough calcium, we get irritated. Our bodies cannot function without proper playmakers, just like a good soccer team.

BUT MORE DOESN'T MEAN BETTER.

It's best to have just a few leaders. Nothing good ever comes from having too many. I will introduce all 17 dietary minerals in this chapter, including how they help our bodies, in which types of food they can be found, and what happens if we take in too much or too little.

---

* There's still some disagreement about which of these are essential to living organisms—some scientists say 13, some say 20 or more. Note that these dietary minerals should not be confused with "minerals" in the general sense, of which there are over 4,000!

minerals are conductors.

Na

CAN BE FOUND IN

pickles

Miso

Dried foods

Soy sauce

sauces

IF YOU DON'T HAVE ENOUGH...

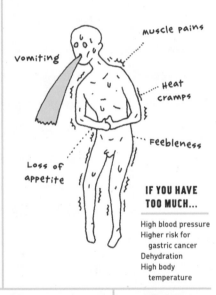

Vomiting

muscle pains

Heat cramps

Feebleness

Loss of appetite

IF YOU HAVE TOO MUCH...

High blood pressure
Higher risk for
    gastric cancer
Dehydration
High body
    temperature

THE MOST IMPORTANT LIFESAVER MINERAL OF THEM ALL

Most of our sodium intake is from table salt (sodium chloride). Many people have cut down on salt in their diet because it can cause problems. But if you ever find yourself sweating a lot or sick with diarrhea, consider taking supplemental sodium because of all the liquid loss, or you might find yourself with a deficiency.

RECOMMENDED DAILY INTAKE (AVERAGE)

600 mg

178

Mg

CAN BE FOUND IN	**MAGNESIUM**	IF YOU DON'T HAVE ENOUGH...

Toasted nori

Spinach

Bananas

kelp

Soybeans

Fish

seaweed   sesame

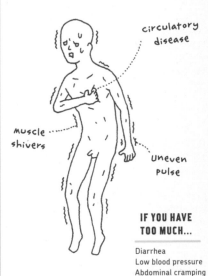

circulatory disease

muscle shivers

uneven pulse

IF YOU HAVE TOO MUCH...

Diarrhea
Low blood pressure
Abdominal cramping

**BUILDING OUR BODIES!
THE MEATY ELEMENT**

Magnesium is found in our bones, where it keeps them strong and helps promote growth, and in our brains, where it helps maintain the thyroid gland. It also helps activate all types of enzymes. Chronic alcoholics should take note: When lots of alcohol leaves our bodies, it takes significant amounts of magnesium with it.

**RECOMMENDED
DAILY INTAKE
(AVERAGE)**

MEN
320 – 370 mg

WOMEN
260 – 290 mg

POTASSIUM

| CAN BE FOUND IN | | IF YOU DON'T HAVE ENOUGH... |

Persimmons

Bananas

sweet potatoes

spinach

Tomatoes

Soybeans

watermelons

sardines

Loss of appetite

Arrhythmia

Respiratory disease

weakness

Vomiting

Diarrhea

muscle paralysis

Hypocalcemia

IF YOU HAVE TOO MUCH...

Hypercalcemia
Cushing's syndrome
Uremia
Urinary occlusion

THE MEGA MULTITASKER

Potassium is always on the move. Be it composing proteins, managing the liquid level between cells, or just taking care of one of the many signaling duties that must be performed, potassium is on the job. Any extra potassium is dealt with by the kidneys, so if they fail, taking too much becomes a definite health risk.

RECOMMENDED DAILY INTAKE (AVERAGE)

MEN
2500 mg

WOMEN
2000 mg

Ca

| CAN BE FOUND IN | **CALCIUM** | IF YOU DON'T HAVE ENOUGH... |

Dairy products

Dried radish

Dried young sardines

seaweeds

Dried shrimp

sardines

Tofu

spinach

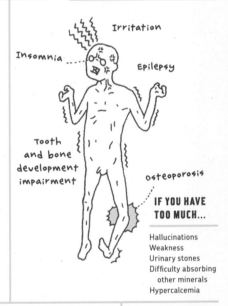

Irritation

Insomnia

Epilepsy

Tooth and bone development impairment

Osteoporosis

IF YOU HAVE TOO MUCH...

Hallucinations
Weakness
Urinary stones
Difficulty absorbing other minerals
Hypercalcemia

THE STEADY MAINSTAY WHO KNOWS HOW TO MAKE STRONG BONES

Most people know that calcium is essential for tooth and bone growth, but its usefulness doesn't stop there, as it has a multitude of other functions. It often works with magnesium, so taking both elements at the same time usually makes them work more efficiently. Vitamin D makes digesting calcium easier.

RECOMMENDED DAILY INTAKE (AVERAGE)

MEN
650 – 800 mg

WOMEN
600 – 650 mg

P

PHOSPHORUS

CAN BE FOUND IN

Dairy products

Seaweeds

Grains

Fruits

Fish and shellfish

Beans

Meats

Nuts

IF YOU DON'T HAVE ENOUGH...

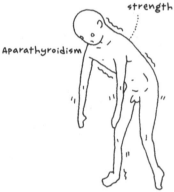

Decreased muscle strength

Aparathyroidism

IF YOU HAVE TOO MUCH...

Calcium absorption difficulties
Hyperparathyroidism
Decreased kidney function

BUILDING OUR DNA!
THE INTELLECTUAL ELEMENT

Phosphorus, famous as the ignition agent of matches, not only is responsible for the information in our DNA but is also a vital component in our cell membranes and neurons. It is also used as an additive in processed foods and as a preservative, so some people think we are taking in too much phosphorus these days.

RECOMMENDED DAILY INTAKE (AVERAGE)

MEN
1000 mg

WOMEN
900 mg

Zn

CAN BE FOUND IN	ZINC	IF YOU DON'T HAVE ENOUGH...

Almonds

Cashews

Oysters

Kouya tofu

Cod roe

Liver

Saury

Scallops

Eel

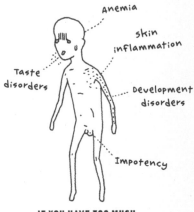

Anemia

Skin inflammation

Taste disorders

Development disorders

Impotency

IF YOU HAVE TOO MUCH...

Gastrointestinal irritation, low blood pressure, uropenia, anemia, pancreatic disorders, increase of LDLs, decrease of HDLs, decrease of immune response, headaches, nausea, stomachache, diarrhea

---

THE LOVING MOTHER ELEMENT

Zinc is required for protein composition as well as correct propagation of gene information and gene expression. Suffering from a zinc deficiency during puberty might affect the development of secondary sex characteristics such as facial hair for men and breast size for women. So even teenagers should eat properly!

RECOMMENDED DAILY INTAKE (AVERAGE)

MEN
11 – 12 mg

WOMEN
9 mg

Cr

CAN BE FOUND IN	CHROMIUM	IF YOU DON'T HAVE ENOUGH...

Black pepper

Whole grains

Brewer's yeast

Beans

Mushrooms

Liver

Shrimp

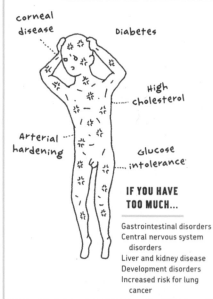

corneal disease

Diabetes

High cholesterol

Arterial hardening

Glucose intolerance

IF YOU HAVE TOO MUCH...

Gastrointestinal disorders
Central nervous system
  disorders
Liver and kidney disease
Development disorders
Increased risk for lung
  cancer

THE GUARDIAN DEITY OF OUR BLOOD SUGAR LEVELS

Most of the chromium in our food is trivalent chromium, which is used in the metabolism of sugars, proteins, and cholesterol. Deficiencies might lead to diabetes or high cholesterol levels, but the amount you need is very small and can be found in basically all foods.

RECOMMENDED DAILY INTAKE (AVERAGE)

MEN
35 – 40 µg

WOMEN
25 – 30 µg

Se

CAN BE FOUND IN

sesame
seeds

Fish and
shellfish

chocolate

Eggs

seaweeds

Beef

Liver

squid

IF YOU DON'T HAVE ENOUGH...

Heart disease

Increased risk of
lifestyle diseases
such as cancer
and Alzheimer's
disease

IF YOU HAVE TOO MUCH...

Fatigue, nausea,
stomachache, diarrhea,
peripheral neuropathy,
liver cirrhosis, rough skin,
hair loss, gastrointestinal
disorders, vomiting, nail
disfigurement

THE YOUNG SUPPORTER, CHEERING LIFE ON

Working as an antioxidant and an immunity booster, selenium helps prevent lifestyle diseases. But having too much is highly toxic and can lead to nail disfigurement and hair loss. It works best when taken together with vitamin E, which can be found in most types of nuts.

RECOMMENDED DAILY INTAKE (AVERAGE)

MEN
30 µg

WOMEN
25 µg

Mo

CAN BE FOUND IN	**MOLYBDENUM**	IF YOU DON'T HAVE ENOUGH...

Liver

Grains

Beans

Dairy products

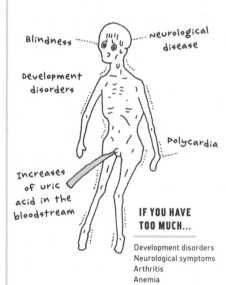

Blindness

Neurological disease

Development disorders

Polycardia

Increases of uric acid in the bloodstream

IF YOU HAVE TOO MUCH...

Development disorders
Neurological symptoms
Arthritis
Anemia

SUPPORTING OUR ENZYMES! THE BODY'S MAINTENANCE MAN

In addition to assisting our enzymes, molybdenum also boosts the effect of iron in our system, which reduces the risk for anemia. We don't need a lot of it, and you should be able to get enough from almost any diet. Milk contains a lot of molybdenum; around 25–75 µg per liter!

RECOMMENDED DAILY INTAKE (AVERAGE)

MEN
25 – 30 µg

WOMEN
20 – 25 µg

Fe

CAN BE FOUND IN	IRON	IF YOU DON'T HAVE ENOUGH...

 soybeans

 chicken

 Liver

 spinach

 Eggs

sardines

 Brown algae

sesame seeds

 Turtle blood

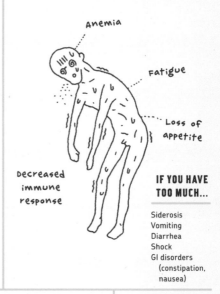

Anemia

Fatigue

Loss of appetite

Decreased immune response

IF YOU HAVE TOO MUCH...

Siderosis
Vomiting
Diarrhea
Shock
GI disorders
(constipation, nausea)

THE LEADER OF THE MINERALS WHO KEEPS US HAPPY AND HEALTHY!

Even the ancient Greeks knew about the relationship between iron and our bodies. Almost 65% of all the iron we consume is used in blood production, so running short is a definite risk. Taking it with vitamin C makes it easier for us to absorb, but tea and coffee have the opposite effect because of something called *tannin*.

RECOMMENDED DAILY INTAKE (AVERAGE)

MEN
7.0 – 7.5 mg

WOMEN
6.0 – 11.0 mg

CAN BE FOUND IN

seaweeds

Fish and shellfish

THE LIFE-FORCE SPOUTING POWER PUMP

IF YOU DON'T HAVE ENOUGH...

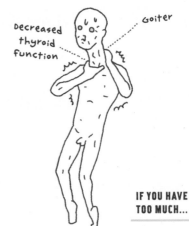

Decreased thyroid function

Goiter

IF YOU HAVE TOO MUCH...

Goiter
Grave's disease
Hyperthyroidism

A mineral that affects both body and mind, iodine is a vital component in the thyroid hormones that control metabolism and the autonomic nervous system. Since it's common in seafood, island nations like Japan have no problem with absorbing enough. Inland areas of America depend on adding iodine to table salt.

RECOMMENDED DAILY INTAKE (AVERAGE)

130 µg

Cu

CAN BE FOUND IN	COPPER	IF YOU DON'T HAVE ENOUGH...

Brewer's yeast

chocolate

shellfish

cow liver

mushrooms

crustaceans

Beans

Fruits

squid and octopus

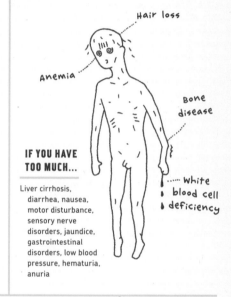

Hair loss

Anemia

Bone disease

White blood cell deficiency

IF YOU HAVE TOO MUCH...

Liver cirrhosis, diarrhea, nausea, motor disturbance, sensory nerve disorders, jaundice, gastrointestinal disorders, low blood pressure, hematuria, anuria

---

STOPPING HEART ATTACKS! THE KEY TO A LONG LIFE

People don't really think of it as a mineral, but there are over 100 mg of copper in an adult body, residing mainly in the blood, brain, liver, and kidneys. It also has a proven preventive effect against heart attacks and arterial sclerosis, so middle-aged and elderly people would do well to eat lots of fish!

RECOMMENDED DAILY INTAKE (AVERAGE)

MEN
0.8 – 0.9 mg

WOMEN
0.7 mg

Mn

CAN BE FOUND IN	MANGANESE	IF YOU DON'T HAVE ENOUGH...

Green tea   seaweeds

Beef   Beans

oysters   Powdered green tea

clams

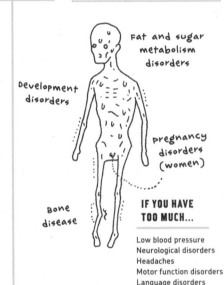

Fat and sugar metabolism disorders

Development disorders

pregnancy disorders (women)

Bone disease

IF YOU HAVE TOO MUCH...

Low blood pressure
Neurological disorders
Headaches
Motor function disorders
Language disorders
Parkinson's-like diseases

THE SUPPORTING ELEMENT THAT NAILS THE IMPORTANT PARTS

A 70 kg adult contains about 12 mg of manganese. It is extra important to pregnant women and affects our motor functions. Experiments with rats have shown that manganese deficiencies can lead to smaller testicles in males. But you don't have to worry about that as long as you have a relatively normal diet.

RECOMMENDED DAILY INTAKE (AVERAGE)

MEN
4.0 mg

WOMEN
3.5 mg

S

Eggs

meats

硫黄
Sulfur

Sulfur is a component of the amino acids that make up the proteins in our bodies and keep us healthy by maintaining our skin, nails, and hair. Deficiencies can lead to skin inflammation and diminished metabolism. It can be found in eggs, meat, and fish.

RECOMMENDED DAILY INTAKE (AVERAGE)

MEN
10 – 12 mg

WOMEN
9 – 10 mg

Cl

soy sauce

Miso

塩素
Chlorine

Chlorine is very important to the digestive system, as it is one of the main components of the hydrochloric acid (gastric acid) secreted into the stomach. As it can be found in table salt, deficiencies should never become a problem. Excess chlorine is excreted through both sweating and urination, so no worries there either.

RECOMMENDED DAILY INTAKE (AVERAGE)

NOT NOTEWORTHY

F

Green tea

Fish and shellfish

フッ素
Fluorine

Fluorine keeps our bones and teeth strong. Since sodium fluoride also has preventive effects on cavities, small amounts are put into the tap water in some areas. Japanese people never have to worry about running low on fluorine since large quantities can be found in both seafood and green tea leaves.

RECOMMENDED DAILY INTAKE (AVERAGE)

NOT NOTEWORTHY

Co

meats

oysters

コバルト
Cobalt

You shouldn't have to worry about cobalt deficiencies if you make sure to eat a lot of seafood and meat proteins, as they contain vitamin B12, which in turn contains the mineral. Not having enough cobalt can lead to anemia, no matter how much iron you take in. It might not be a very versatile element, but it is important nonetheless.

RECOMMENDED DAILY INTAKE (AVERAGE)

NOT NOTEWORTHY

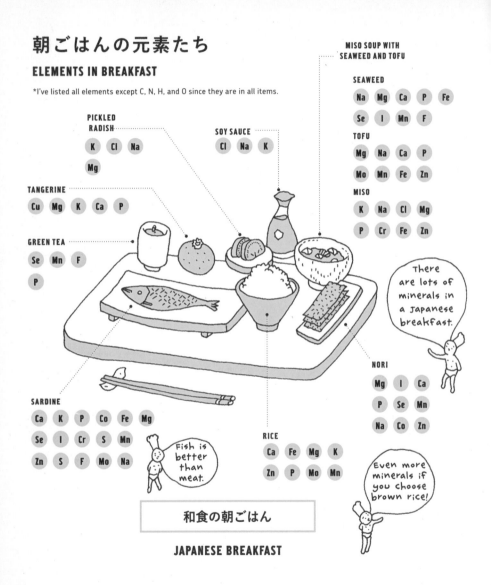

朝ごはんの元素たち
ELEMENTS IN BREAKFAST

*I've listed all elements except C, N, H, and O since they are in all items.

和食の朝ごはん

JAPANESE BREAKFAST

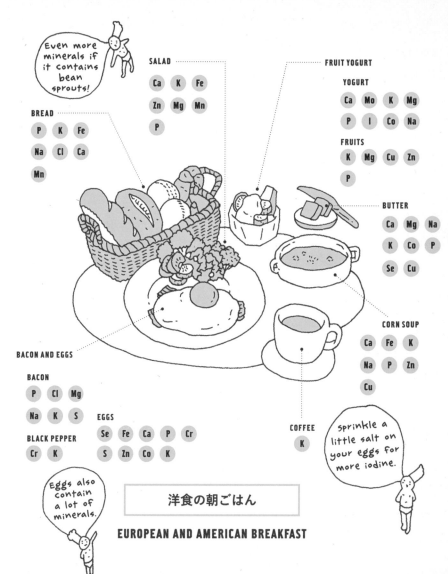

洋食の朝ごはん

EUROPEAN AND AMERICAN BREAKFAST

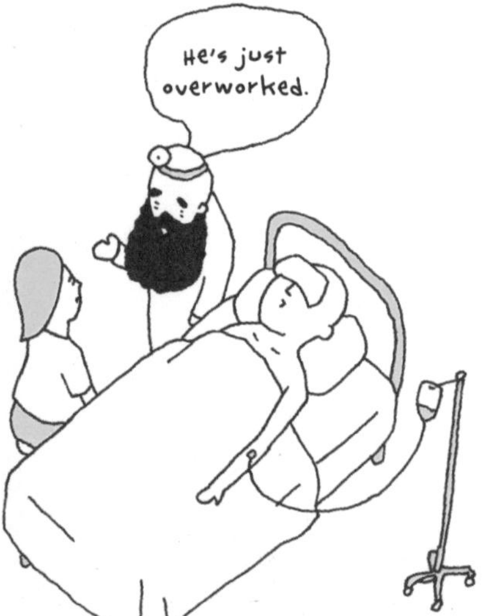
He's just overworked.

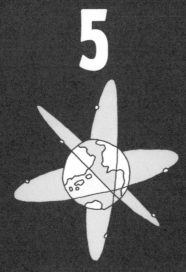

5

THE ELEMENTS CRISIS

元素危機

Some of the elements we've looked at so far, like germanium, were very popular a few years back but aren't really used any more. Other elements like indium only recently came into the spotlight.

SOME ELEMENTS ARE SO POPULAR, IT'S BECOMING A PROBLEM.

Long ago, batteries were made using nickel. Because of this, the price of nickel skyrocketed, forcing us to come up with the lithium battery as a cheaper replacement. Indium, used in LCD displays, is also getting more expensive by the year. Scarce elements like indium and elements that are generally very hard to process or extract are called *rare metals*.

ALMOST ALL RARE METALS IN JAPAN ARE IMPORTED TODAY.

Of course, Japan didn't really have any natural rare metal resources to begin with. Since Japan is importing almost its entire demand for rare metals, it would be extremely bad if that stream of raw materials were ever to stop.

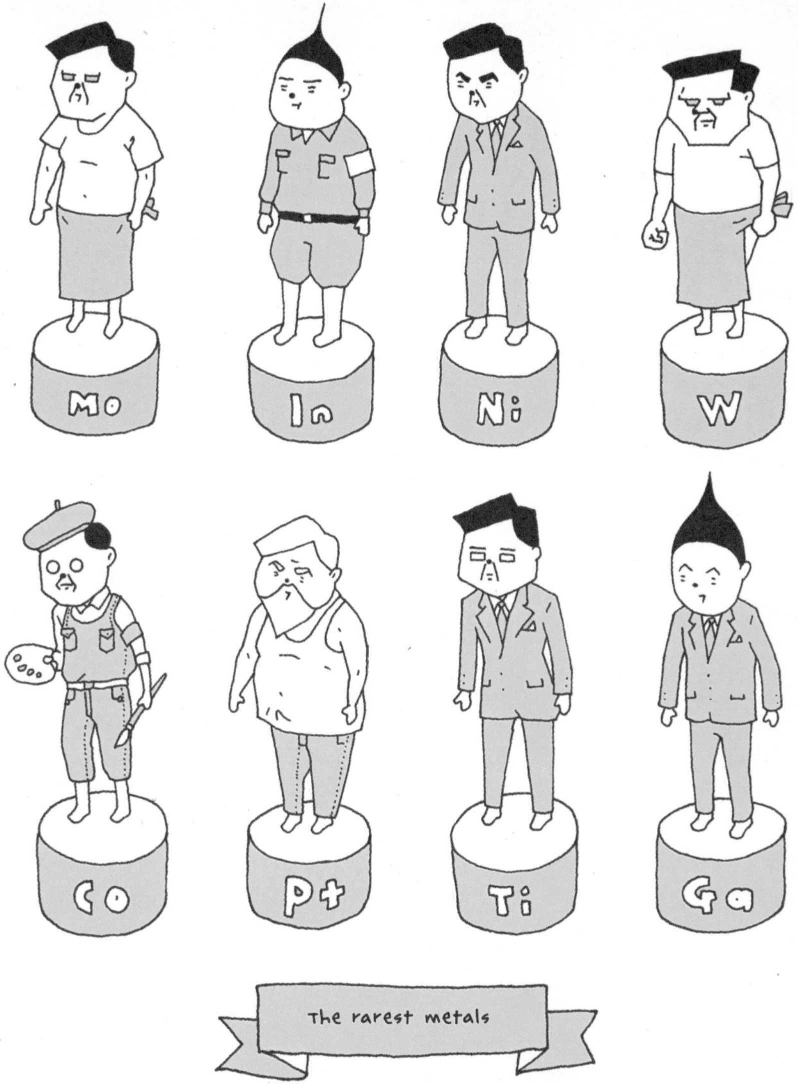
Mo
In
Ni
W
Co
Pt
Ti
Ga
The rarest metals

Tungsten is required to make the tools we need to build things. Nickel and molybdenum imports let us create stainless steel products. And gallium and its related metals are the basis for our semiconductors. No semiconductors means no computers or mobile phones. These few elements carry Japan's economy on their shoulders.

BUT THE RISK OF AN ELEMENT CRISIS IS VERY REAL.

The popularity of some metals has driven their price up to the point that it's hard to acquire them at all. This is true not only for Japan but for the entire world. This makes the element crisis at least as serious as the impending oil crisis, and some countries have already begun stockpiling hard-to-find elements while they promote research for potential replacements.

But it might not be enough. We, as different countries and cultures, must learn to work together to solve the crisis.

we won't be able to
make strong tools
without tungsten.

The manufacturing
industry comes to a stop.

we can't make
semiconductors when
elements like gallium
run out.

This also means no more
computers and other high-
tech equipment.

LCD TVs
require
indium.

stainless steel
is made of
molybdenum and
nickel.

And batteries are
made with lithium.

We are now able to perform advanced recycling of home electronics and even mobile phones. It's not just about being kind to the environment, it's also about reclaiming precious rare metals from our garbage. In some cases the element could become unrecoverable if not processed correctly.

WE CANNOT MAKE ELEMENTS.

Why don't we just make elements if we need them so badly? Just put two hydrogen atoms together and you've got helium! The protons and electrons are all there, so how hard can it be?

IF WE COULD CREATE THEM LIKE THAT, THEY WOULDN'T BE ELEMENTS.

An atomic reaction or an incredible amount of energy is required to reshape an atomic nucleus. But inducing atomic reactions produces radioactive materials, which emit dangerous radioactive rays. The elements are called elements because they are hard to create and alter.

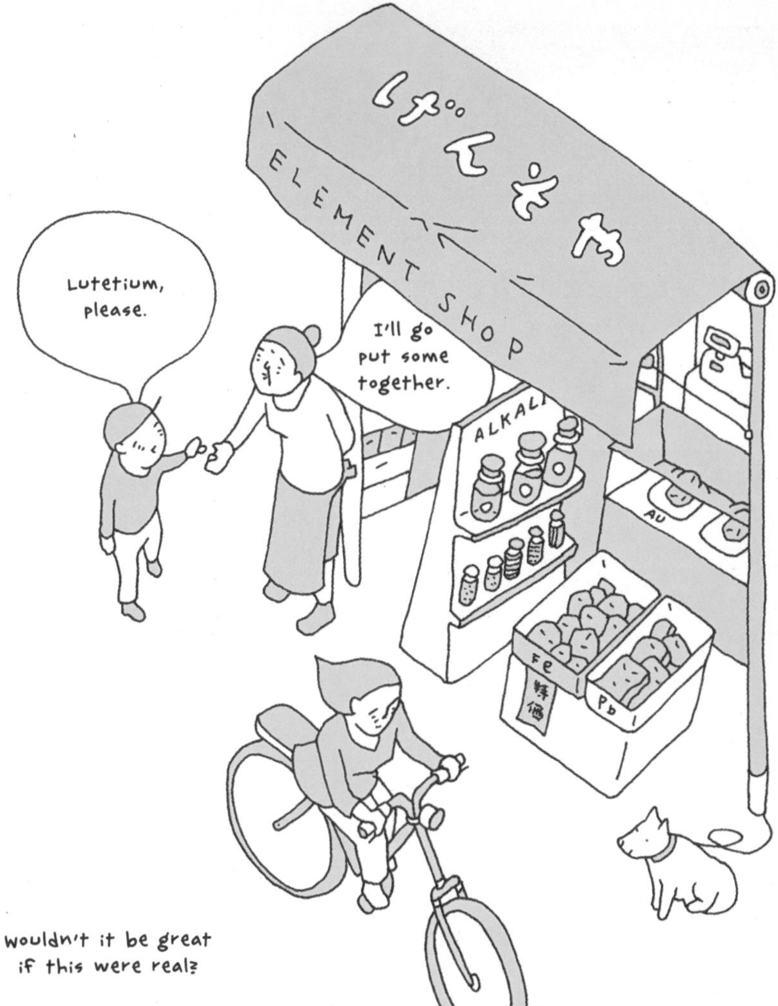
Lutetium, please.
I'll go put some together.
げんそや
ELEMENT SHOP
ALKALI
AU
Fe
特価
Pb
wouldn't it be great if this were real?

Our current way of life is supported by our use and knowledge of elements. It might not be apparent, but elements are responsible for the most basic parts of our modern world.

IN THE FUTURE, EVERYONE WILL BE A SCIENTIST.

The concept of the "low-carbon economy" has become more popular lately. Maybe we need to start examining our environmental problems at the element level as well. The greenhouse gas problem, for example, is aggravated by us humans releasing underground carbon dioxide into the atmosphere. The element crisis is of course another problem, and I'm hoping that you will become more aware of your rare metal usage after getting to know these elements a little better.

If we could get everyone to take an interest in the elements that make up our world and apply that knowledge in their daily lives, this looming crisis may never come to pass. I would be honored if you decided to adopt a more rare metal–aware lifestyle after reading this book.

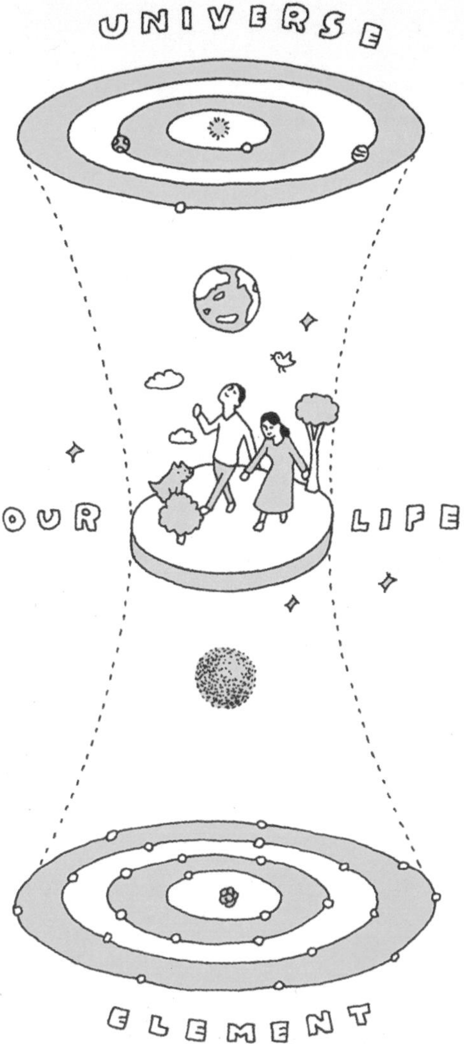
UNIVERSE
OUR LIFE
ELEMENT

INDEX

AFTERWORD

I imagine many people remember which element they first heard about. Mine was uranium. I was still in primary school when I saw the movie *Barefoot Gen* with my mother at the local community center. As some of you may know, the movie is about the bombing of Hiroshima during World War II. I still remember the intensity of the movie, and by the end of the show, it had rendered my young self completely speechless. The following weeks I had trouble sleeping, and the scene where the bomb explodes haunted me day and night. I convinced myself that I had to learn more about the bomb, not because I had some passing interest in it, but because I felt that I would never be able to let it go if I didn't. I was completely terrified. It was then that I first learned of the elements uranium and plutonium and of the world of neutrons, protons, and electrons. I recall how calming it was to read about the bomb and how it worked.

When I was contacted by Fumiko Kakoi of Kagaku Doujin to make a book about the periodic table, I didn't think much of the idea at first. I didn't really know much about the elements, even after my illuminating (and traumatic) experience with *Barefoot Gen* as a child. I wasn't sure how to proceed but finally decided to meet with Professor Kouhei Tamao of the Institute of Physical and Chemical Research and Professor Hiromu Sakurai of Kyoto Pharmaceutical University. They taught me about the impending element crisis and about the importance of the metals present in our bodies. It was a truly eye-opening experience to hear about the intricate bond that our bodies share with the elements. Everything I learned there and from then on finally coalesced into the book you're reading right now. I would like nothing more than to let my old self, the one who didn't care about the elements, read it, and I hope that it can be of help to anyone else who might want to take a gander.

I didn't complete this book by myself—far from it. My little sister Makiko Kajitani, who also happens to be a writer, helped me so much in so many ways that it might have been more fair to list her as a co-author. I am also very grateful to Takahito Terashima, whom I sadly never met, who helped me greatly in editing the book. And my companion for two years now, Kakoi-san of Kagaku Doujin, has helped me with every aspect of the book, from research and gathering materials to proofreading. Words cannot adequately describe the gratitude I feel toward you all.

Thank you so much.

Bunpei Yorifuji

ABOUT THE AUTHOR

Born in 1973 in Nagano, Japan, Bunpei Yorifuji is a
Musashino Art University dropout. His other books
include *The Catalog of Death* (*Shi ni Katarogu*) and
The Scale of Mind (*Suuji no Monosashi*). He has also
co-authored *Uncocoro* and *The Earthquake Checklist*
(*Jishin Itsumonooto*), among others. Find out more
about Bunpei and his works at *http://bunpei.com/*.

THE PRODUCTION TEAM FOR THE
JAPANESE EDITION

Book Design: Bunpei Yorifuji and Ayaka Kitatani
Publisher: Ryosuke Sone
Publishing House: Kagaku Dojin

THE PRODUCTION TEAM FOR THE
ENGLISH EDITION

Publisher: William Pollock
Production Editor: Serena Yang
Developmental Editor: Keith Fancher
Translator: Fredrik Lindh
Technical Reviewer: Brandon Budde
Compositor: Riley Hoffman
Proofreader: Paula L. Fleming

UPDATES

Visit *http://nostarch.com/wle* for updates, errata,
and other information.